Synergies for Sustainable Energy

For a complete listing of titles in the
Artech House Power Engineering Series,
turn to the back of this book.

First and foremost, to my loving parents, thank you for always supporting and encouraging me in everything that I do and shaping me into the person I am today.

To all my friends and colleagues, who I've learned so much from and who have always been by my side.

To this nurturing planet we inhabit, full of energy, resources, and all things we need but often misused.

And to future generations, in hopes that we can leave a healthy planet for you to live wonderful lives on.

Contents

Preface

In the current literature, there are many books that examine various aspects of the clean energy arena. Some go into the details of a specific technology (or group of technologies), some discuss policy issues, some explain the underlying science, some go back to the history of those technologies, and some discuss the potential transitions to a sustainable future. While there is a wealth of information out there—and each resource serves a purpose—a resource to connect all the dots and explain the underlying synergies is needed.

Energy issues are better understood with better solutions potentially developed if viewed from a wider perspective. This wider perspective would display all the synergies, as well as the causes and effects within the technology marketplace and development phases. When issues are considered in isolation, important problems—as well as key opportunities for valuable synergies—are missed. Connections are not always easily and clearly made when matters are discussed separately.

This book has been conceived to tie all the pieces together in an attempt to explain the current status of various technologies and their potential future challenges and developments. It begins with an overview of related key global trends, followed by a focus on providing a snapshot of global energy issues (such as the status of resources and consumption trends), and then ties them to the related social, historical, and environmental developments. The reader is then guided through the various clean energy technologies, energy-saving methods, and energy storage applications, starting with key facts and the technology basics. We then delve into the main players, markets, and relevant analyses and their results, as well as challenges and opportunities. The concluding chapter connects the pieces by explaining what this all means at the local/regional/global scales.

A diverse range of readers may benefit from this book (such as engineers, managers, and decision-makers from industry, utilities, international organizations, federal/state/local governments, cities/regions, and the clean energy investing community). Technologies and related issues are presented in such a way that a more complete and accurate understanding may be formed, thus enabling better decisions to be made. Often, without this synergistic and comprehensive view, decision-makers can over- or underestimate the costs and consequences resulting from their decisions. This can result in missed opportunities to leverage resources, costly decisions due to misconceptions, or missing the most efficient applications to overcome technology hurdles.

This book was written with the goal of empowering decision-makers in filtering through the deluge of complex energy-related issues and information with a renewed, streamlined, more systematic, and synergistic perspective.

1

Introduction

1.1 Key Global Trends

When we look back at history and examples of collapsing human civilizations, it can be seen that humans often do not anticipate upcoming negative trends, failing to appropriately take action to manage these threats. Extensive research [1] has identified 12 emerging global trends that may pose significant challenges for humankind in the years ahead. These trends are briefly explained below, in no specific order. There are inevitable interactions between some of these trends, and energy is a significant component.

1. *Repercussions of "The Crisis":* How to deal with global public debt levels of around $39 trillion is a major issue, and new economic, financial, political, legal, and social structures are expected to shape the 21st century.
2. *Geopolitical Power Shifts:* The BRIC (Brazil, Russia, India, and China) economies are on a rapidly developing path, led by China. China's GDP has already surpassed the economies of Western European countries, and is expected to approach that of the United States by 2035. India, Russia, and Brazil will follow in second, third, and fourth place, respectively. These countries are also expected to catch-up with European and U.S. economies within the 2020 to 2040 timeframe.
3. *Technological Challenges:* While further incremental changes to current technologies will continue, eventually these improvements will diminish. When these technologies reach maturity, they will inevitably become vulnerable to new, innovative technologies that will have

the capability to change societies, similar to the way the personal automobile, air transport, and the Internet impacted and transformed societies in the 20th century.

4. *Climate Change:* There's strong underlying scientific evidence on climate change, though debate continues. However, the issue has reached a point that now it is beyond a matter of mitigation, and will also require adaptation. While humans need to find serious solutions to reducing greenhouse gas emissions, they will also need to find ways to adapt.
5. *Water and Food Supply:* Water is becoming the next precious resource. We will be facing increasing scarcity if this resource is not managed well. Some countries and regions are more susceptible to the decreasing levels of groundwater supplies, as their economies are more dependent on this resource. Worldwide, approximately 70 percent of water is used in agriculture. This not only signifies the imbalance in resource use, but diminishing water supplies also could end-up meaning less food to feed the growing global population.
6. *Education:* Though there have been advances Asia, Africa, and Latin America continue to fall behind in educational development. Globally, over 16 percent of the population cannot read.
7. *Demographic Changes:* The 21st century will bring demographic challenges with an ever increasing global population that challenges the planet, though the population level is predicted to eventually stabilize.
8. *War, Terrorism, and Social Unrest:* Human history has always been witness to violence and social problems. While the 20th century closed with a sum of many wars and millions of casualties, the 21st century has already started off unstable, and the instability may incease.
9. *Energy Supply:* Energy is the main driving force behind all human activity. Energy demand keeps growing, especially with the increasing development of the BRIC nations. This increasing demand will continue to place a burden on fossil-based energy sources, which already have a big share in global energy consumption. The share of renewables in the global energy mix is growing and this growth is expected to continue, indicating an area for further development, which has the potential to bring positive change.
10. *Ecosystems and Biodiversity:* Humans are leaving a large ecological footprint with their wasteful living habits, and are on a path to use-up the equivalent of two Earths to support their needs by 2050. There has already been a great decline in species—nearly 40 percent less today—compared to 1970.

11. *Health:* While many major illnesses and health concerns of the past that took away millions in populations have been eradicated, different parts of the global population still suffer from poor health conditions. In poor nations, a considerable number of people still suffer and lose their lives from common, easy to cure diseases. On the other hand, factors such as the emergence of drug-resistant microorganisms, new child illnesses, and ailments brought about by urban lifestyles are bringing new health challenges to the more developed nations.
12. *Natural Disasters:* The combination of populated cities, poor urban planning, insecure construction practices, damage to natural buffers, and climate change have made the world's population even more vulnerable to the impacts of natural disasters. At the same time, phenomena like the global economic crisis and increasing geopolitical tensions among countries add to the global population's vulnerability and decreased ability to cope with major shocks and negative impacts.

1.2 Snapshot of Global Energy Trends

The following subsections provide an overview of global energy trends, looking at various resources used for energy, the consumption of energy, and carbon dioxide emissions resulting from these activities.

1.2.1 Energy Resources and Production Pathways

The primary energy supply of the world has been based on fossil resources. While renewable resources are inching their way into the energy picture, their overall share is still very small. In 1973, the global energy mix was composed of 86.6 percent fossils. About 3.5 decades later, in 2009, the percentage of fossils decreased slightly to 80.9 percent, while the main increase was observed in nuclear, along with some increase in hydroelectric and other renewables. Despite this trend, the share of all renewable resources (hydro, biofuels and waste, geothermal, solar, wind, heat, etc.) in 2009 was only 13.3 percent, still a very small fraction of the global energy supply, and demonstrating only a 6 percent increase over 3.5 decades [2].

Looking at the primary global energy supply from a different perspective, it can be seen that the majority of this supply (61.3 percent) came from OECD countries in 1973. In 2009, while OECD countries still accounted for a large portion (43.3 percent), over the 3.5 decades, other regions and countries started to play a more noticeable role. Sizeable increases in supply can be seen in China (from 7 to 18.7 percent), the rest of Asia (5.6 to 12 percent), and the Middle East (0.8 to 4.8 percent) [2].

Similar to the global trends in primary energy production, the global production of electricity continues to be dominated by fossil resources. While coal maintains its dominance amongst fossil resources used for electricity generation, compared to the mid-1970s, oil's share in electricity production has significantly declined (from 24.7 percent in 1973 to 5.1 percent in 2009), while natural gas' share has increased (from 12.1 percent in 1973 to 21.4 percent in 2009). While the share of renewables for electricity production has increased, it still only represented 3.3 percent of global production in 2009 [2].

1.2.1.1 Fossil-Based Production: Oil

Crude oil production has been dominated mainly by the Middle East (30.4 percent in 2010), followed by OECD countries (21.7 percent in 2010) [2].

One issue that inevitably comes-up in debates relating to fossil-based energy use patterns is the peak oil debate. In the mid-1950s, former U.S. Geological Survey and Shell Oil geophysicist, M. King Hubbert, made a prediction —which turned-out to be pretty accurate—that U.S. domestic oil production would peak in 1970, and decline from that point on, never reaching that peak point again, following a bell curve (where the area under Hubbert's bell curve represents the ultimate recoverable quantity of the resource). He also indicated that this peak in production would take place sometime after a peak in the reserves (quantity of known resource that is economically recoverable at the current day's prices) had been reached. In the United States, this lag time was 11 years. The reserves peaked in 1959 and the production peaked in 2000. Hubbert made similar forecasts for global oil production, claiming that it would peak around the year 2000 [3].

According to the International Energy Agency (IEA) [4], by 2035, U.S. oil imports are expected to decline by about 50 percent (due to rising domestic output and improvements in transport efficiency), while European Union (EU) oil imports are expected to surpass those of the United States, and China is predicted to become the largest oil importer.

Predictions of the timing of peak oil on a global basis are not clear cut. Though the predictions don't exactly match, they are still relatively close, and more importantly, the overall trend is the same. We are pretty much on the downward path on the slope. The exact timing of the global peak in oil production depends on various factors, such as recoverable resources, demand, alternative and more efficient technologies replacing oil-based ones, and OPEC production decisions [5]. Predictions from IEA over the next couple of decades demonstrate that after about 2009, current crude oil fields have started to be supplemented by yet to be *developed* oil fields. This delays peak for just a couple years, and then follows the same declining trend. At the same time, there are three other potential future resources: yet to be *found* crude oil, natural gas liquids, and unconventional oil. While these three potential resources would help considerably with adding to the resource base, apart from the fact that they also are finite, they will become harder and more costly to exploit [4].

1.2.1.2 Fossil-Based Production: Natural Gas

While natural gas production had mainly been dominated by OECD countries and non-OECD Europe and Eurasia in the mid-1970s, over the last couple decades the OECD countries' share has beenreduced by about 50 percent. Various other regions of the world (such as the Middle East and Africa) have started to produce significantly larger amounts of natural gas [2].

By 2035, the top three natural gas producers are expected to be Russia, United States, and China. The discovery and utilization of unconventional natural gas supplies will be the main source of these rankings for the United States and China. Unconventional natural gas supplies are also expected to contribute to most of Canada and India's supplies. Overall, unconventional natural gas is expected to provide around 40 percent of the global supply increase, which in turn will bring an increased need for stricter environmental controls [6].

1.2.1.3 Fossil-Based Production: Coal

The global production of coal has not only increased, but also demonstrated a pretty significant change in the players involved. While back in 1973, OECD countries accounted for half of the global coal production, by 2010, China became the main coal producer accounting for half of the global production. At the same time, since the beginning of the 21st century, world coal production has displayed a steady increase, due mainly to increased production in China [2].

1.2.1.4 Nuclear-Based Production

Production of global nuclear energy increased steadily from the mid-1980s through the beginning of the 21st century. Since then nuclear energy has remained fairly steady, and even on a slightly downward trend. Production has been dominated by OECD countries [2].

Global nuclear energy production increased at a faster rate from the early 1970s until about the 1990s, but between then and the beginning of the 21st century, growth has slowed down. In fact, the global share of nuclear power compared to total electricity production has been stable at around 17 percent over the last decade or so, even declining over the past few years [7]. The March 11, 2011 earthquake that resulted in the Fukushima nuclear plant disaster in Japan has further altered the nuclear landscape, causing Japan and other countries (Germany, Italy, Switzerland) to start considering phasing-out nuclear power. Thus, the years ahead might display a further decline in nuclear energy's global share.

1.2.1.5 Renewables-Based Production: Hydroelectric

Hydroelectric energy production worldwide has experienced an upward trend since 2005, especially with the increased production in China and Latin America. While a significant part of production (71.9 percent) was in OECD countries back in 1973, by 2009, these countries' share had decreased to 41.7 percent, with China and Latin America gaining significant market shares, at 18.1 percent and 20.1 percent, respectively [2].

1.2.1.6 Renewables-Based Production: Wind

Global installed wind power capacity has been growing, and reached 237,016 MWs in 2011 (which is about 30-fold of the amount installed about a decade and a half ago), supplying around three percent of the global electricity demand in 2011 [8].

1.2.1.7 Renewables-Based Production: Solar

Global cumulative installed photo voltaic capacity has grown approximately five-fold between 2006 and 2010. While eight main countries contribute to this installed capacity, Germany has taken the lead, approaching nearly 50 percent of global installed capacity [9].

1.2.2 Energy Demand and Consumption

Currently, per capita energy use around the world displays quite a wide range (two order of magnitude) from the deficiencies and poor living conditions of sub-Saharan Africa, to the over-consumption of wealthy societies in the United States, Europe, and Asia where even the average per capita energy use in some of the most prosperous societies reaches levels far beyond what is needed for comfortable living) [10].

The consumption of energy globally is on an upward trend, with the consumption of oil dominating the landscape. While OECD nations used to be the majority consumers at 60.3 percent in 1973, in 2009, China, the rest of Asia, and the Middle East grew to be major consumers as well [2].

Worldwide primary energy demand is predicted to be driven mainly by emerging economies, with global energy demand increasing by approximately one third between 2010 and 2035. China and India will account for 50 percent of this growth [6].

By 2035, natural gas and renewables are expected to meet about two-thirds of the global energy demand (Figure 1.1) [6].

Currently, there are close to 800 million cars traveling the roads around the world, and this number is expected to increase to two to three billion by 2050. Transportation is also the most rapidly growing source of *greenhouse gas* (GHG) emissions, with emissions from transport expected to increase 250

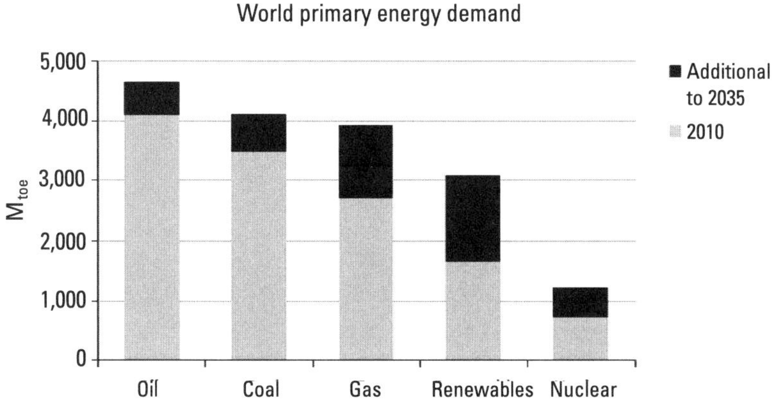

Figure 1.1 Resources needed to meet global primary energy demand by 2035 [6].

percent by 2050, mostly in developing countries [11]. The global passenger vehicle fleet is expected to approximately double to 1.7 billion by 2035, and most of this growth is expected in non-OECD countries, especially China and India [6]. Demand for transportation fuels will increase corresondingly. Considering the dynamics of imports and exports of oil around the globe, the current main consumers of this resource are the United States, Europe, Japan, and China, while the main producers are the Middle East, the former Soviet Union, Africa, and Latin American countries [1]. If the demand for transportation fuels is still dependent on oil, resource and environmental issues will be exacerbated. Whereas, if alternative fuels are able to increasingly meet this demand, the world will have started to travel towards a more sustainable path.

1.2.3 Carbon Dioxide Emissions

Our fossil fuel-dependent energy patterns have been leading to excessive GHG emissions. About 80 percent of the global GHG emissions are due to our energy-related activities. Compared to the preindustrial times, atmospheric concentrations of carbon dioxide have increased by 30 percent [12].

Another environmental phenomenon of global nature is the depletion of the ozone layer. This problem was able to garner more cooperation from nations for solutions as it was easier to find substitute chemicals for the damaging ones. However, the global warming issue is a much more complex one. It involves far more extensive changes to an established and functioning energy infrastructure that has served humanity for a long time. The global response has been a mix of reactions. From opposing parties at the individual to the country level claiming that changes would cause economic harm, to supporters leading the way in developing new technologies, taking initiatives, and making strong commitments. At the same time, the *connected* nature of everything in our ecosystem shows us again that fixing damage after it is done is not as easy. One example is the interrelationship between ozone depleting chemicals and global warming-causing GHGs. Both the chemicals that damage the ozone layer and the substitutes developed are also strong GHGs. So, while the efforts to phase-out ozone depleting chemicals helped address climate change, some of the substitutes created (such as HCFCs) *contributed* to climate change. Moreover, increasing climate change is actually *slowing* the ozone layer's recovery. Climate change is warming the troposphere (atmosphere near the Earth's surface), but cooling the stratosphere (where the ozone hole damage has occurred), where a cooler stratosphere means conditions that hasten the ozone-depleting chemical reactions [12].

Following the trend of fossil use throughout the world, global CO_2 emissions are also on the rise, with OECD countries contributing a large portion of the emissions, while China's CO_2 emissions have realized a significant increase from 5.7 percent in 1973 to 23.7 percent in 2009 [2].

Energy-related CO_2 emissions are expected to remain a challenge over the next few decades. From the present to 2035, cumulative global CO_2 emissions are expected to surpass three quarters of the total since 1900, while China's per-capita CO_2 emissions are expected to match the OECD average. As can be seen from Figure 1.2, while CO_2 emissions between 2010 and 2035 have a reduced share (compared to emissions between 1900 and 2009) in the United States, European Union, and Japan, the opposite situation is expected to hold true for China and India [6].

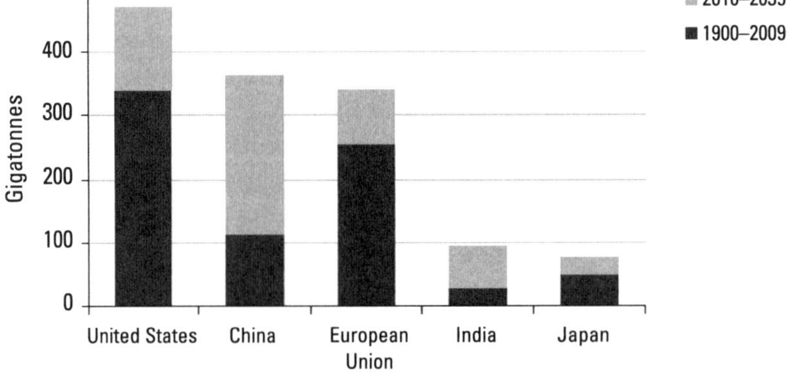

Figure 1.2 Cumulative energy-related carbon dioxide emissions (by country)—past and projected [6].

One alarming fact about the CO_2 emissions is that even if we were to immediately reduce or even stop all emissions, the effects would linger on for quite extended timeframes, because of the slow removal period of CO_2 once it has accumulated in the atmosphere. Even if CO_2 emissions were to peak then decline over the next 100 years, it would take a couple centuries for CO_2 and the temperature to stabilize, while the effects of sea level rise due to thermal expansion and ice melting would continue anywhere from a few centuries to a few millennia [3].

1.3 Quest for Sustainability

Sustainability may be defined as "patterns of economic, environmental, and social progress that meets the needs of the present day without reducing the capacity to meet future need." Consequently, *sustainable energy* indicates: "patterns of energy production and use that can support a society's present and future needs with the least life-cycle economic, environmental, and social costs." In order to be considered truly sustainable, our actions and our choices should not exclude options, nor place environmental and economic burdens on future generations. Moreover, any decision and/or action related to energy needs to be considered from a life-cycle basis, to ensure consideration of all related aspects. Only then can options be truly compared relative to their advantages and disadvantages Moreover, the life-cycle benefits and costs need to consider economic, environmental, social, and security aspects, as each has a unique and important role to play [3]. Sustainability can only be accomplished if unity is achieved on a small and increasingly crowded planet, where "the winners will lose if the losers do not win" [11].

While events like the global economic recession and the nuclear disaster in Fukishima, Japan have been reflected in short- and medium-term energy trends, the overall and long-term trend of increasing energy consumption continues. Global economic turmoil and stress has caused attention to shift away from developing energy policies, and towards dealing with the immediate economic concerns. Some current global trends do not paint a picture of a world headed in the right direction. CO_2 emissions have rebounded to a record high; it's the second straight year that the energy efficiency of the global economy has worsened; and spending on oil imports have reached record highs. Although the global energy landscape is complex, dynamic, and uncertain, the overall trend of emerging economies and increasing population will continue to drive increasing energy needs. While the global diversity of oil supply is shrinking, natural gas is experiencing a "rebirth" through the exploration of unconventional sources, and the future of coal will be dependent on the use of more efficient power plants and the use of carbon capture and storage technologies [6].

1.4 Population Growth: History

Around 2,000 years ago, at the first millennium A.D., the world's population was around 300 million people. A thousand years passed by, and the population increased by only 10 million. It continued to increase at slow rates, less than 0.1 percent per year. However, the Industrial Revolution of the late 1700s was a tipping point, resulting in great technological advances, which raised living standards and thus prompted growth. By 1800, global population had reached one of its key milestones—one billion people. Of this billion, most were concentrated in Asia (65 percent), 21 percent were in developing Europe, and less than one percent were in North America. Within the first half of the 20th century, between 1920 and 1950, the growth rate of global population was still relatively low—around one percent per year. However, it was still more than that experienced in the previous centuries, and just 127 years after the milestone in 1800, in 1927, the global population reached two billion. From mid-20th century on, with advances in medicine, agriculture, and sanitation in the developing countries, life expectancy and birth rates increased. Thus, only 33 years after the two billion mark, in 1960, the global population reached three billion. By the latter part of 1960, the population growth rate had increased to 2.04 percent per year. On an ever shortening time scale, this time only 14 years later, in 1974, world population reached four billion. Most of the global population increase was observed in the developing countries. The global population reached five billion and six billion by 1987 and 1999, respectively. Currently (similar to the situation back in 1800) the majority of global population resides in Asia [13].

Global population reached seven billion on October 31, 2011 [14], and is projected to reach 9.3 billion by 2050 and 10.1 billion by 2100. The expectation is that the majority of the population increase will be in Asia, followed by Africa. However, different scenarios also lay out higher or lower population possibilities. The United Nations (UN) has indicated that the global population in 2050 could be as low as 8.1 billion and as high as 10.6 billion, while in 2100, it can range between 6.2 billion and 15.8 billion [15].

On a more granular level, each region will experience population increases and declines at a different rate. Of the two regions of the world with the largest contributions to population growth, Asia's growth has reached the zone (five-year period between 2010 and 2015) where decline is happening. However, Africa's population is still expected to expand, until around 2050–2055, when it is expected to demonstrate a slower path to decline [15].

Moreover, the overall global population growth has experienced and exponential growth since around the middle of the 20th century (Figure 1.3) [13].

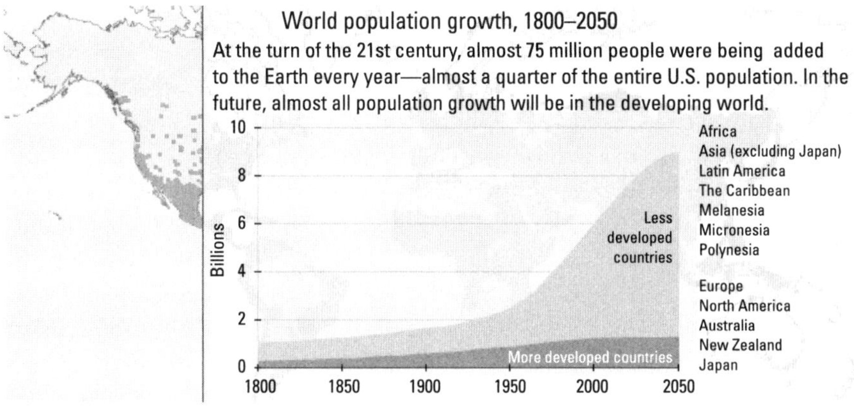

Figure 1.3 Exponential growth of population, 1800–2050 [13].

1.5 Population Growth: Impacts

During 1970–1972 the Systems Dynamics Group of the Sloan School of anagement at the Massachusetts Institute of Technology utilized system dynamics theory and computer modeling to analyze the growth of the world's population, along with causes and consequences [16]. The group has published several analysis results: *Limits to Growth* (1972), *Beyond the Limits* (1992), and *Limits to Growth: The 30-Year Update* (2004). The original *Limits to Growth* conveyed the key point that global ecological constraints were expected to have great effects on global developments in the 21st century, and also cautioned that a significant amount of capital and manpower would most likely be required to deal with these constraints, in effect causing the average quality of life to decline sometime in the 21st century. The end of growth was predicted to be about 50 years after the publication of the book in the early 1970s, providing hope that there was adequate time to consider issues, make better choices, and take corrective actions. If growth leads to overshoot (which is the increase in demands on the planet's resources) and sinks that are beyond sustainable levels, this situation will lead to collapse. In 1972, when the original analyses were made, it still seemed like there was room to grow in a safe manner, while having a chance to examine longer-term options. However, in just two decades, that assumption did not hold true anymore. The study results published in 1992 reinforced the original findings, while, more importantly, also pointed out that humans had already overshot the carrying capacity of the Earth.

Currently, our world is experiencing a great deal of stress from various pressure points caused by a combination of population growth, superfluous consumption, waste, and unsustainable energy production and consumption patterns. While there are also significant advances being made in science and technology, resulting in more efficient, clean and advanced technologies, the decisions and actions taken today will determine the ultimate fate of upcoming generations, but also of the quality of life we experience today.

There is an imbalance and troublesome trend in that material and energy consumption per capita in the richest parts of the world are exceeding sustainable levels, while the poorest 1.3 billion people continue their need to consume in order to break away from their poverty [17]. However, if developing nations follow the consumption patterns of the developed nations, our already-stressed planet will not be able to handle the burden. Even though there is an overall decline in the rate of population growth throughout the world, the world's population continues to grow, at 80 million per year, and the highest fertility rates are seen mainly in the least developed countries (exacerbating the consumption situation) [17]. The population-consumption-environment correlation is a dynamic and complex one.

There are many trends involved with regards to growth in population and industry. These trends usually reinforce or are in conflict with each other. As people's wealth improves, they demand more industrial products, but also want reduced pollution levels. The energy and materials needed to maintain the growth of industry are also depleting finite resource stocks and deteriorating renewable resources. At the same time, there is progress in creating technologies that are able to use materials more efficiently and to enable the discovery of new reserves [16].

1.6 The Earth's Carrying Capacity

The relationship between humans and the supporting ecosystems of the planet is a dynamic one. Analysts have determined four different possibilities that population and the Earth's carrying capacity can develop and interact with each other, as shown in Figure 1.4 [16, 17]. In the most optimistic case (a) population and carrying capacity do not come into conflict (the planet's limits are either far away or are growing faster than population). The second case (b) can also be considered optimistic, as the population growth slows down as it reaches the carrying capacity limit. However, neither of these scenarios is available to the global population anymore, as the global society has already far surpassed limits. The third case (c) is one where society overshoots its carrying capacity, but does not result in big or permanent damage, as there is continuous adjustment taking place and the ecological footprint would oscillate near the limit before leveling off. The fourth case (d) is the most troubling one, where the limits are surpassed to an extent that the damage caused is severe and permanent,

How population can approach its carrying capacity

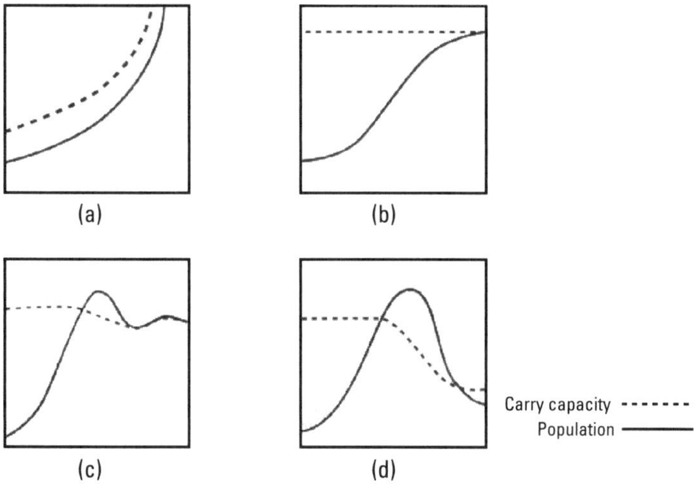

Figure 1.4 Population and the Earth's carrying capacity [16, 17].

with environmental collapse and resource decline. The global population and economy would rapidly decline to reach a balance with the new, much lower carrying capacity.

1.7 Our Ecological Footprint

The world is currently experiencing heightened ecological stress. The unprecedented scale and pace of resources being pushed beyond their limits is leading to this stress, which ends up triggering loss of species, deforestation, and a decrease in the ability of the planet to absorb increasing amounts of waste. Since the Industrial Revolution, the model followed by mankind has been to grow towards production and consumption levels that are unsustainable, growing economies at any cost, instead of developing economies that allow for ecological restoration. It has been estimated that we are currently consuming 1.5 of Earth's ecological capacity [11].

Energy and materials flow from their planetary *sources* to their eventual *sinks*. However, limits exists regarding the rates at which sources are able to produce, and sinks are able to absorb, without eventually hitting a point where harm is done to people, the economy, and the planet's natural mechanisms of regeneration. Regarding fossil-based energy resources, either the source or the sink may become the limiting factor. For example, for coal, the current quantity available (the source) is still quite large, but its use will most likely be limited by the atmospheric sink for carbon dioxide generated from its use. Oil on the other hand, may be limited at either the source (first fossil resource likely to be depleted) or the sink (as it also deals with carbon dioxide and other pollution issues) [16].

Until the 1980s, global environmental problems were not necessarily regarded as major issues by the world's governments. However, with the increase in threats to our planet, as well as the rise of environmental movements, global environmental issues have taken a more prominent place in global politics. Environmental issues are closely linked with other issues such as economic development, international trade, conflicts, and national stability. Alterations to the atmosphere, oceans, soil cover, plants, animals, and the climate systems caused by humankind throughout history have led to growing concerns. While human populations increased and prospered with economic development, the byproducts of these developments (GHG emissions, release of ozone-damaging substances, water pollution, toxic chemicals production, exploitation of natural resources, reduction of forest cover) have placed cumulative pressures on our planet, threatening the health and well-being of humans. The product of the world's population, per capita consumption, and waste production defines humanity's stress on the planet, or its *ecological footprint*. The World Wildlife Fund evaluates global ecological footprint as "humanity's demand on the biosphere in terms of the area of biologically productive land and sea needed to provide resources demanded, as well as to absorb the wastes created" [12]. As an example, in 2005, global demand was already 30 percent more than the supply of the world's total biocapacity. Moreover, for over 50 years, the per capita consumption of natural resources has been increasing much faster than the rate of population growth. Humanity's ecological footprint initially surpassed the planet's biocapacity in the 1980s, and as developing nations increase in population and develop to levels demanding lifestyles similar to developed countries, more stress will be placed on the planet. The United States holds less than five percent of the world's population, but consumes approximately a quarter of the global fossil fuel resources [12].

In 2005, the single largest demand that human consumption put on the biosphere was its *carbon footprint*, which had already reached levels ten times higher than in 1961. The United States and China were the main contributors to this carbon footprint, with each using 21 percent of the world's biocapacity, followed by India [12]. At present, while the world's richest countries comprise only a small fraction of the global population, they account for more than half of the global energy use. This trend points to troubling implications for the future. If developing nations, with their much larger (and increasing) share of global population continue down the path of the developed nations, the stress put on our planet will result in very serious consequences.

Some key energy and environmental milestones, issues, and international developments from 1800 to the present day are highlighted below [12]:

- *1800:* The carbon dioxide and methane levels in the atmosphere are around 270–290 parts per million (ppm) and 700 parts per billion (ppb). These levels are used today as the pre-Industrial Revolution baseline numbers in comparisons and analyses.
- *1827:* The first speculating about the existence of the greenhouse effect was published by French mathematician and physicist Jean-Baptiste Joseph Fourier.
- *1859:* Irish physicist John Tyndall becomes one of the first scientists to study the greenhouse effect.
- *1896:* An article is published by Swedish scientist Svante Arrhenius, claiming that doubling the amount of carbon dioxide in the atmosphere would lead to temperatures rising by 5°C to 6°C.
- *1900:* The carbon dioxide concentration in the atmosphere reaches 295 ppm.
- *1945:* The United Nations is established.
- *1952:* A mix of dense fog and sooty, black coal smoke leads to the death of 4,000 to 12,000 people in London.
- *1982:* Formal negotiations on the protection of the ozone layer begin.
- *1985:* The Vienna Convention for Protection of the Ozone Layer is signed. The discovery of the Antarctic ozone hole is published in *Nature.* Climate experts at a conference produce consensus on the serious possibility of global warming.
- *1986:* As a result of a major explosion at the Soviet nuclear plant in Chernobyl, radioactivity spreads across Western Europe and Japan.
- *1987:* The Montreal Protocol on Substances that Deplete the Ozone Layer is signed.
- *1988:* Several governments agree to form the Intergovernmental Panel on Climate Change (IPCC).
- *1989:* Oil spills from the *Exxon Valdez* in the Gulf of Alaska. The Noordwijk Declaration is issued in the Netherlands, calling for the stabilization of carbon dioxide emissions in the atmosphere by 2000.
- *1990:* The first assessment report of the IPCC is released, stressing that the global surface temperature has increased by 0.3°C to 0.6°C during the past 10 years.
- *1992:* The United Nations Conference on Environment and Development is assembled in Rio de Janeiro, Brazil and the United Nations Framework Convention on Climate Change (UNFCCC) is signed.

- *1996:* The second assessment report of the IPCC is released, which states that the influence of humans on global climate is evident.
- *1997:* The Kyoto Protocol to the Framework Convention on Climate Change is signed.
- *2000:* The Millennium Summit is held in New York, at the United Nations headquarters.
- *2001:* The third assessment report of the IPCC is released, this time with a more alarming message that the influence of humans on global climate is stronger than ever.
- *2002:* The United Nations World Summit on Sustainable Development is held in Johannesburg, South Africa, on the 10th anniversary of the signing of the UNFCCC.
- *2004:* The concentration of carbon dioxide in the atmosphere reaches an alarming amount of 379 ppm.
- *2005:* The Kyoto Protocol enters into force. The first multicountry, multisector GHG emissions trading scheme—the European Union's Greenhouse Gas Emissions Trading Scheme (EU ETS)—begins operation. Data indicate that 2005 was the warmest year since regular records were kept more than a century ago.
- *2006:* It is reported by NASA that the recovery of the ozone layer has been greater due in part to the reduced chlorofluorocarbons (CFCs) phased-out by the Montreal Protocol.
- *2007:* The fourth assessment report of the IPCC is released, further confirming that climate change is in fact occurring, that impacts due to humans are unequivocal, and that these impacts will increase as temperatures keep increasing. The European Parliament calls for the reduction of GHGs to limit the increase in global temperatures to only 2°C by the year 2100. The United States is left as the only major industrialized country not having ratified the Kyoto Protocol.
- *2008:* Carbon dioxide levels in the atmosphere reach 387 ppm, the highest levels observed in the last 650,000 years. The European Union announces that it intends to reduce its GHGs by 20 percent from 1990 levels by the year 2020 and that it will require that 20 percent of the Union's total energy consumption come from renewable energy sources. In the western part of Antarctica, a 160 square mile piece of the glacial ice collapses.
- *2012:* Within just four days, nearly all of Greenland's ice sheet melted, with the ice melt area going from 40 percent of the ice sheet to 97 per-

cent (as compared to 55 percent realized as the greatest melt of the last three decades) [18].

1.8 Civilization and Energy

Energy is a critical part of human development and civilization. Throughout the ages, humans have found ways to make use of the variety of resources of the planet they inhabit. They did so by initially farming plants and domesticating animals, and then moving on to finding more complex approaches to provide more power through the development of machines that utilized energy resources. The discovery and use of energy resources and the progress of civilizations are intimately linked. Each phase of human development has been triggered by changes in energy use. Since the middle of the 19th century, energy use has supported a fourfold increase in global human population and the global economy to grow fortyfold [3].

A very close relationship exists between the growth of civilization and patterns of energy use. Approximately 100,000 years ago, mankind discovered that rubbing two stones together could create a fire. This meant thermal energy, and is believed to be the first human-engineered conversion of energy. From 8000 B.C. to 4000 B.C., mechanical energy was the primary source, with the discovery of the wheel and stone tools, as well as the domestication of animals for work. Use of thermal energy obtained initially from wood fire, and then from coal emerged between 4000 B.C. and 1000 B.C., providing warmth, cooking, and assisting in developing materials such as pottery, bronze, and iron. In the first century A.D., the power of water and wind were used for grist milling and water pumping. Also starting in A.D. 100, coal and oils were initially used to provide for heat and lighting, then later used in the development and expansion of industry. These resources were utilized as fuel for the steam engine of the 1800s, and then for other heat and mechanical engines, paving the way for more streamlined industrial processes and improved mobility. Key discoveries and developments then led to the fossil fuel based land and air transport widely used and accepted today: the first commercial oil well in 1859; invention of the internal combustion engine in 1877; discovery of oil in Texas, U.S., in 1901; invention of the airplane in 1903; discovery of oil in Iran in 1908; and the introduction of assembly line production of cars, as well as the affordable Model T by Ford. The invention of the generator and motor in 1831 signaled the beginning of the prominence of electric power. From 1950 on, further growth of the use of fossil fuels, electricity, and related technologies in various sectors of the economy led to unforeseen population and economic growth [3].

Figure 1.5 displays the trend of energy consumption by source for the United States [20]. It can be seen over time that resource use follows the trend

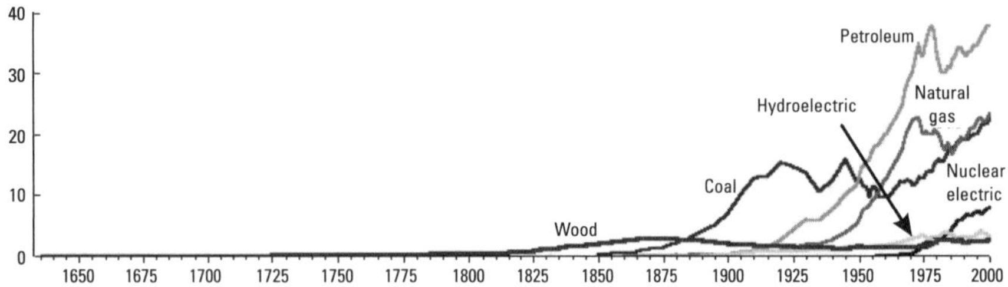

Figure 1.5 Energy consumption by source in the United States, 1650–2000 [20].

of global civilization, with activity starting to pick up in the 1800s, and then rapidly growing, with a major reliance on fossil resources.

1.9 Energy Throughout History and Transitions

Energy has been and will continue to be central to sustaining life and human development , enhancing welfare and improving societies. It has been shown throughout history that every major change in living standards has, at its core, been tied to a reform in in energy use [1]. Energy has enhanced human development from providing for basic needs such as heating and cooking, to the development of streamlined industrial processes, faster transportation, and modern gadgets that put the world at the tip of our fingers. Humans started with lower energy consumption expectations, relying mostly on manpower and simple forms of utilizing the power of nature. Until the Industrial Revolution, humans relied on plants as a source of food and fuel, and on animals as a source of food and mechanical energy [19]. However, the discovery of the vast amounts of buried fossil resources resulted in an explosion of energy use with corresponding improvements, while simultaneously resulting in the rapid degradation of the very nature that we had initially depended on for our energy and well-being.

Once large deposits of coal were discovered and made technologically accessible (starting near the coal fields of Britain), industrial activity increased rapidly, and spread to Europe, North America, and the rest of the world. Coal was an extremely concentrated form of energy, and allowed for steam production, which was the driving force behind factories, trains, and steamships. Thus, access to coal meant rapid economic development. Then, the curiosity and efforts of mankind led to the discovery of yet another, more energy-packed fossil resource—oil—under the ground in the late 1800s and early 1900s. Oil proved to be more convenient to transport (in pipelines) and led to a multitude of end-products, from transportation fuels to a host of chemicals and materials. The development of the first electricity-generating units in the 1880s allowed for greater reliance on discovered fossil fuels, a trend spreading rapidly throughout the globe [1].

The energy arena poses multifaceted challenges to a range of stakeholders from the individual, to businesses, to countries, and societies as a whole. The 21st century has revealed the issue of *energy security*. To be able to develop and sustain a modern economy will inevitably require nations to have access to vast amounts of *reliable* energy. Nations need to use their own local energy resources, and/or have access to an affordable international market to provide for their energy security and drive their economic development [1]. With world dependency on fossil fuels increasing energy security and reliability poses more challenges. These fossil resources are unevenly distributed across the globe are diminishing. While renewable resources may not necessarily be abundant everywhere, each country has its unique set of renewable resources. Some parts of the world, like island nations, have been blessed by richer resources, but that certainly does not mean that those are the only areas that can benefit from these resources. With the right technologies and policies, renewable resources have the potential to help nations ranging from making them 100 percent renewables-based to at least reducing the difficulties of fossil-dependence. The key here is that these resources are renewable, and, even if not always in the greatest quantities, are still available, for the most part, in every location. Look at it this way, is there a location without sunshine? The wind also always blows (maybe more in specific locations) but, it would be harder to find a country with very poor wind conditions throughout. How about waste-to-energy, a concept gaining more popularity, especially in the highly-populated urban areas? As long as there are humans, there will be waste generated.

Transitions in energy fall in a continuum between sudden changes and trends that are gradually revealed. For energy substitution to take place, new infrastructures need to be developed, while at the same time, existing capital investments need to be discarded. This process is a slow one, typically requiring the new energy source and technology approximately a century to gain half of market share. Until the 1970s, this process timeframe was thought of as regular; it was later observed that the pace and timing of these transitions were much less predictable than previously thought, with events such as oil price hikes disrupting systematic energy substitutions [21].

The prevailing converters of energy are still highly dependent on fossil resources, thus slowing down the evolution of modern energy systems. Some of the main examples are the steam turbine and the internal combustion engine. The steam turbine is a key energy converter that was invented about 120 years ago, and is a high-load prime mover that converts 70 percent of our energy in fossil- and nuclear-based facilities. The gasoline-based internal combustion engine was introduced at the same time as the steam turbine, reached maturity in just one generation, and remains the core prime mover for transportation all around the globe. Further complicating the shift from fossil-based resources to renewables are factors such as the scale of the shift required, as well as the lower energy density and intermittency of renewables [21].

Energy systems started with independent and small-scale units. Systems in the pre-industrial age were batches of independent units, with limited spatial coverage (e.g., a small village sustaining its fuel and food needs from nearby forests and crop residues). As societies started to develop and modernize, the boundaries of energy systems also began to expand, starting with transport of coal with rail and ships, to national and international trade. At present, our energy system can be considered truly global, with approximately 50 countries exporting and nearly 150 countries importing crude oil (with a similar trade in refined oil products), over 20 countries trading natural gas, close to a dozen coal import-export activities occurring, and about two dozen countries having established interconnections with their neighbors to exchange electricity on a GW scale [10].

Historically global energy transitions have been defined by inertia, small adjustments, and slowly developing innovations in the pre-industrial era, to rapid development, spread and adoption of innovations after industrialization. A time span of approximately five millennia during the pre-industrial age was signified by the dominance of burned biomass resources for heating (households, metallurgy, other manufacturing), and the use of human and animal muscles as the source of mechanical energy (with the exception of sails). Even in the early modern era when some Western European societies started to extract peat and to deploy more powerful water wheels and windmills the dominance of biomass-based energy was not really affected. The biggest energy transitions in our history have actually happened in a relatively short timeframe. The switch from the use of biomass to fossil fuels, as well as the shift from reliance on human or animal power to the power of machine, has only happened in the last few centuries, or even just a few decades, while the advent of electricity began as recently as the 1880s. Although the shift from biomass-based resources to fossil fuels has in general followed the patter of a move from complete reliance on fuel wood and charcoal to increasing uses of coal, there have been exceptions in some regions and countries. For example, between 1608 and 1672, the Dutch Republic realized a shift from biofuels to peat (youngest form of fossil fuels), and was also supplemented by prevalent use of wind power. Another example during the 20th century, is when several countries in Africa and Asia (with no domestic coal resources) shifted from biofuel use to using refined oil products and natural gas. Some desert countries even became the largest users (on a per capita basis) of hydrocarbons, within just a matter of two generations. However, though there are exceptions, the main global pattern (especially in the major economies of the world) has been the typical and gradual shift from biofuels to coal [10].

1.10 The Three "Revolutions"

History has unveiled two significant revolutions (*agricultural* followed by the *industrial)* that drastically changed the landscape of civilizations. Approximately 10,000 years ago, increasing populations throughout the world who had lived lives of nomadic hunter-gatherers, started realizing that their growing numbers were surpassing the plants and animals they used for food. To cope with this situation, some chose to migrate to areas of the world still rich in plants and animals. Others chose to stay in one place, cultivating plants and domesticating animals. This was a new concept that had profound effects. These settled farmers were able to obtain more food per hectare than their fellow hunter-gatherers. However, the nutritional quality of the food was lower, there was less variety, and more effort was required in production. As opposed to nomads, the settled farmers became more vulnerable to factors such as weather, diseases, pests, and invasions from others. Wastes and pollution started becoming a problem. While the development of agriculture was a solution for dealing with food source scarcity, it also allowed the populations to expand at rapid rates, thus creating new scarcities of land and energy. It was time for the industrial revolution which started in England with abundant reserves of coal replacing the use of dwindling trees. While the use of coal initially resulted in various problems (earthmoving, construction of mines, pumping water, transport, etc.), solutions to these problems were found relatively quickly and technology started taking an increasingly important place in society. The industrial revolution led to significant improvements and developments in societies, but also brought with it population growth, pollution of our natural resources, and increased energy use. Similar to the agricultural revolution that preceded it, the wide-spread success of the industrial revolution gave rise to scarcity; this time not only with regards to fuels, materials and land, but now also relating to the carrying capacity of the planet. The next revolution (sustainability) that awaits mankind is already underway. It will take time to unfold, and will also lead to significant changes that, while being hard to fathom, are expected to drive mankind onto a better path [16].

References

[1] Done, A., *Global Trends: Facing Up to a Changing World,* Palgrave Macmillan, 2012.

[2] International Energy Agency, *Key World Energy Statistics,* Paris, France: International Energy Agency, 2011.

[3] Randolph, J., and G. M. Masters, *Energy for Sustainability: Technology, Planning, Policy,* Washington, D.C.: Island Press, 2008.

[4] International Energy Agency, *World Energy Outlook: 2011 (Executive Summary),* Paris, France: International Energy Agency, 2011.

[5] Website: The Oil Drum, "Discussions About Energy and Our Future." Available at: http://www.theoildrum.com/.

[6] International Energy Agency, "World Energy Outlook 2011," Presentation to the Press, London, United Kingdom, November 9, 2011.

[7] Website: World Nuclear Association, "Nuclear Power in the World Today." Available at: http://www.world-nuclear.org/info/inf01.html.

[8] *World Wind Energy Association, 2011 Report,* Bonn, Germany: World Wind Energy Association, May 2012.

[9] U.S. Department of Energy, *2010 Solar Technologies Market Report,* Washington, D.C.: U.S. Department of Energy, November 2011.

[10] Smil, V., Energy *Transitions: History, Requirements, Prospects,* Santa Barbara, CA: Praeger, 2010.

[11] Worldwatch Institute, *State of the World 2012,* Washington, D.C.: Worldwatch Institute, 2012.

[12] Chasek, P. S., D. L. Downie, and J. W. Brown, *Global Environmental Politics,* Boulder, CO: Westview Press, 2010.

[13] Website: NOVA Science Programming on Air and Online, "World in the Balance: Human Numbers through Time," Available at: http://www.pbs.org/wgbh/nova/worldbalance/numb-nf.html.

[14] Website: United Nations News Centre, "As World Passed 7 Billion Milestone, UN Urges Action to Meet Key Challenges," October 31, 2011, Available at: http://www.un.org/apps/news/story.asp?NewsID=40257

[15] United Nations, *Department of Economic and Social Affairs, Population Division, World Population 2010 (Wall Chart),* ST/ESA/SER.A/307, New York, NY: United Nations, 2011.

[16] Meadows, D., J. Randers, and D. Meadows, *Limits to Growth: The 30-Year Update,* White River Junction, VT: Chelsea Green Publishing Company, 2004.

[17] United Nations Population Fund, *The State of World Population 2011,* New York, NY: United Nations Population Fund, 2011.

[18] Borenstein, S., "NASA: Strange and Sudden Massive Melt in Greenland," Yahoo! News, July 24, 2012, Available at: http://news.yahoo.com/nasa-strange-sudden-massive-melt-greenland-193426302.html.

[19] Livi-Bacci, M., *A Concise History of World Population (Fifth Ed.),* West Sussex, UK: Wiley-Blackwell, 2012.

[20] Energy Information Administration, *Annual Energy Review 2000,* DOE/EIA-0384(2000), Washington, D.C.: Energy Information Administration, August 2001.

[21] Smil, V., *Global Catastrophes and Trends: The Next 50 Years,* Cambridge, MA: The MIT Press, 2008.

2

Energy Efficiency

2.1 Basics

Implementing energy efficiency measures is a worthwhile goal, as it is the "low-hanging fruit," and should also be part of a portfolio of actions. In trying to meet growing energy needs, energy efficiency represents being able to obtain more value from the energy used and using less of it while maintaining the comfort and benefits energy offers. Energy efficiency is a lower-cost option to meet rising energy demands, with a focus on reducing wasted energy. This cuts consumption, peak demand, and pollutant emissions, while sidestepping needs for upgrades to generation capacity. Programs for energy efficiency may be implemented in a short time frame, typically within one to three years. The energy efficiency measures implemented culminate in benefits to consumers, the implementing organizations, and the nation as a whole, in terms of economic, environmental, and energy security. Implementing energy efficiency measures allows consumers to control their energy use, gain a better understanding of the way they use energy, and identify areas for improvement. Moreover, businesses and industry are also able to gain a competitive advantage through these investments. While investments in energy efficiency provide immediate to short-term benefits, these investments also result in influencing the future cost of energy, since investing in energy supply costs several times more than investing in efficiency.

Increasing the efficiency of buildings, industry, and vehicles results in reduced energy consumption. This translates into less fossil fuel use, reduced harmful emissions, relief in the electricity grid, and protection against disruptions in energy supply and price. While these results may also be attained through the development and use of alternative energy sources and technologies, energy efficiency is a more easily achievable goal. While development and implementation of alternative technologies may take years, various energy efficiency measures can be executed in a much shorter timeframe. However, this does not mean that alternative technologies should not be pursued and that efficiency only will take us fully down the road to sustainability. Efficiency will need to be part of a portfolio of goals.

The fuel crisis in the United States in the 1970s peaked people's interest and attention in energy conservation and efficiency measures for a time. Energy efficiency incentives were implemented and started becoming wide-spread in the early 1980s. As more energy supplies became available, attention was turned to the competitive market, with the prevailing belief that the competitive market would offer energy efficiency services and that there was no need for regulatory measures. However, today the energy and environment-related concerns are more complex and urgent, necessitating a need to revisit energy efficiency measures as part of a portfolio of actions. Investment in energy efficiency is not consistent and equal among the states. A couple of states represent a majority of this funding. Energy efficiency-related efforts that have been sustained in California for over three decades have demonstrated how per capita energy consumption may remain relatively flat, while consumption in the United States as a whole has grown [1].

2.1.1 Federal, State, and Local Support

There are several programs at the federal, state, and local level in the United States that assist communities in increasing energy efficiency.

2.1.2 Weatherization Assistance Program

In partnership with states and over 900 local agencies the DOE's Weatherization Assistance Program provides weatherization services to 100,000 low-income families every year, producing savings of 30.5 million Btu annually. DOE typically provides 40 percent of the weatherization funding, while states, the U.S. Department of Health and Human Services Low-Income Home Energy Assistance Program, and utilities contribute the remaining funds. The Weatherization Assistance Program implements cost-effective energy efficiency measures tailored to each home's specific needs and characteristics. The reduced energy expenditures are about $437, approximately 35 percent reduction in heating costs [2].

2.1.3 State Energy Program (SEP)

The State Energy Program (SEP) provides grants and technical assistance to states and U.S. territories to encourage energy conservation and to control energy demand, typically saving over $300 million in energy costs in a year. The SEP also supports strategic partnerships with various parties, such as governors, state and utility policymakers, local government leaders, and private industry. Additionally, "special projects" are cofunded through competitive state solicitations [2].

2.1.4 Energy Efficiency and Conservation Block Grant (EECBG)

The Energy Efficiency and Conservation Block Grant (EECBG) Program was funded for the first time by the American Recovery and Reinvestment Act (Recovery Act) of 2009, authorized in Title V, Subtitle E of the Energy Independence and Security Act (EISA) of 2007. It aims to support U.S. cities, counties, states, territories, and Indian tribes to develop, promote, implement, and manage energy efficiency and conservation projects. The goal is to empower local communities to make strategic investments to meet the nation's long-term goals for energy independence and leadership on climate change [3]. Grants in the amount of $3.2 billion have been provided to over 2,300 U.S. cities, counties, states, territories, and Indian tribes. This funding supports a wide range of activities: energy audits and energy efficiency retrofits in residential and commercial buildings; advanced building codes; financial incentive programs; transportation programs that conserve energy; energy efficient traffic signals and street lights; combined heat and power and district heating and cooling systems; and others [2].

Activities eligible for use of funds include the following [4]:

- Development of an energy efficiency and conservation strategy.
- Building energy audits and retrofits, including weatherization.
- Financial incentive programs for energy efficiency such as energy savings performance contracting, on-bill financing, and revolving loan funds.
- Transportation programs to conserve energy and support renewable fuel infrastructure.
- Building code development, implementation, and inspections.
- Installation of distributed energy technologies including combined heat and power and district heating and cooling systems.
- Material conservation programs including source reduction, recycling, and recycled content procurement programs.

- Reduction and capture of greenhouse gas emissions generated by landfills or similar waste-related sources.
- Installation of energy efficient traffic signals and street lighting.
- Installation of renewable energy technologies on government buildings.

Any other appropriate activity that meets the purposes of the program and is approved by DOE.

2.1.5 Better Buildings, Better Plants Program

An effort in the United States that brings together industry and the government in implementing voluntary actions to drive down energy intensity in buildings and industrial facilities is the Better Buildings, Better Plants Program. This initiative is led by the Advanced Manufacturing Office (AMO) and the Office of Energy Efficiency and Renewable Energy (EERE) of the U.S. DOE. Companies *pledge* to reduce the energy intensity of their industrial operations by 25 percent or more within 10 years. DOE supports companies that take a pledge through several actions: assigning a technical account manager to analyze key energy use data and metrics, assisting them with learning more about effective energy management, providing a variety of tools and trainings, and providing opportunities for recognition [5].

2.1.6 Local Community Support

Local governments and community organizations may serve supportive and important roles in implementing energy efficiency measures. The more targeted and specific policies that local governments are able to implement can meet that region's energy needs, address community priorities, and also drive demand for utility energy efficiency programs. In the United States, the convention in most states has been for utilities to be the main providers of energy efficiency services. While utilities have become effective at this role, there is still room for improvement and a role for other entities. As more local players enter the field of energy efficiency, the opportunity for these players to work in partnership with utilities emerges, as each has its relative strengths. When coordinated, these groups can enable greater energy savings. In many communities, the local governments have close ties with the communities in which they serve, and also have a direct influence on policies to be implemented in their communities. Nongovernmental local organizations (such as community groups) may not have as direct an impact on policies, but they are in a unique position to leverage relationships with the community to benefit outreach development efforts. In fact, other community-related benefits usually emerge from implementing energy efficiency measures and these additional benefits are usually the drivers for local players. One of the benefits that concerns localities is job creation. If local employees can be matched-up with utility efforts, it becomes a win-win situation. In some cases local jurisdictions adopt energy codes and standards that are more strict than those required at the state or federal level, thus driving improvements. Additionally, in the energy arena, permitting issues are frequently a roadblock to timely project implementation. If the local authorities have a positive influence and involvement, issues related to this potential major roadblock may be resolved [6].

2.2 Efficient Buildings

In the United States, homes and commercial buildings are the main consumers of electricity. Additionally, summer peak electricity demands that pose a burden on the transmission and distribution systems are also mainly associated with buildings, driven by air conditioning needs. Homes and commercial buildings are also dominant in terms of natural gas consumption [7]. Energy is used in buildings for providing cooling, heating, water heating, and lighting. Designing and retrofitting buildings with better windows, more insulation, tighter building envelopes, and ducts, all help manage heat losses reducing energy demand. Larger commercial buildings have a smaller surface/volume ratio, so heat loss through the building envelope per unit floor space is not as significant as in residential units. Office buildings tend to use more energy for lighting, and also have the added heat load that equipment (computers, copiers, etc.) and humans give off.

There are several market barriers to implementing energy efficiency measures in buildings. These improvements are difficult to measure and not clearly visible, especially if there isn't any independent verification of the energy savings achieved. Firms in the buildings industry are relatively small, so it is difficult for them to absorb the related costs and risks. The ownership structure of residential and commercial buildings also has an impact on the execution of energy efficiency programs. Building occupants do not have much incentive to perform efficiency improvements, as they are usually not the owners. Building owners also do not have an incentive to implement energy efficiency programs, as they are not the ones paying the utility bills. Also, the buildings-related professions: architects, developers, construction companies, engineers, and energy service providers are not organized and do not work synergistically to align their strategies related to siting, construction, operations, and maintenance activities [7].

Building designs have evolved over the past few decades, influencing energy demands and corresponding energy management and efficiency considerations. In the 1960s to the 1970s, the main consideration was the building "envelope" and thus the heating efficiencies. In the 1980s, as air conditioners started making their way into markets, in addition to heating-related concerns, energy use by air conditioners started being factored-in. By the end of the 20th century, between the 1990s and 2000, the "whole building" was the predominant concept, where, in addition to heating and cooling, the use of appliances and lighting operating efficiencies were added to the list of considerations. With the 2000s, these considerations expanded even further to include life-cycle environmental and health impacts, as well as life-cycle "embodied energy"[1] in the building materials. Currently, since around 2010, the goal is that of the "whole community," where in addition to previous considerations, attention is now being given to matters such as onsite energy generation, site/neighborhood design, and regional connectivity [8].

Buildings play an important role in energy consumption, with their considerable energy, carbon and environmental footprints. Applying energy efficiency measures to buildings helps reduce energy footprints, while at the same time providing economic benefits to the residents and building owners. In 2010, the U.S. buildings sector accounted for 7 percent of global and 41 percent of U.S. primary energy consumption. The energy consumption of buildings in the United States was 44 percent more than the transportation sector and 36 percent more than the industrial sector. Of the energy consumed in the buildings sector in the United States, homes accounted for 54 percent, while commercial buildings accounted for 46 percent, with most of the energy used coming from fossil fuels. The four main energy consumption end uses in buildings are space heating (37 percent), water heating (12 percent), space cooling (10 percent), and lighting (9 percent) [9].

Primary energy consumption in the United States by buildings has demonstrated a steady increase (48 percent), between the years 1980 and 2009, and is projected to stagnate for a while (due to the effects of the economic recession). Consumption will then go on a steady growth trend from 2016 to 2035. This increase in energy consumption in buildings is closely related to the increases in population, number of households, and commercial floor space, all of which are expected to rise by approximately 30 percent by 2035 [9].

While the U.S. industry's use of electricity has remained relatively flat over the last two decades, the use of electricity in buildings has shown a significant upsurge, leading to an increased burden on power plants. As electricity use in the United States is dominated by the use of coal, this translates into increased fossil resource use and carbon emissions [10].

While there are driving forces (saving money, reducing carbon, organizational drive, etc.) behind investing in energy efficiency for new construction (or including these measures in renovations), there are also regulations and rules

1. Embodied energy is the energy required to manufacture building materials, transport these materials, and construct the buildings.

like building codes that also play a role. There are several trends emerging with regards to energy codes for buildings. Codes are becoming stricter, more states and municipalities are adopting these strict codes, and compliance and enforcement are improving. In fact, some communities are even voluntarily implementing codes that go beyond current requirements. For example, the energy code for commercial construction in Massachusetts is 20 percent stricter than the state's baseline code [11].

Just as consumers planning on buying a car would like to know what the vehicle's fuel economy is (as increasing gasoline prices are a major contributor to the cost of car ownership), those planning on buying or renting a building are also interested in the energy performance of that building. As a response to this consumer demand, *mandatory disclosure regulations (or commercial benchmarking laws)* have been issued in some markets in the United States and various locations throughout the world. These regulations require building owners to reveal data related to the energy consumption of their property. This provides a solution to the "split incentive" issue the building sector faces. Building owners typically have not had an incentive to invest in energy-saving measures, as the tenants are responsible for paying energy-related costs. With the execution of the mandatory disclosure regulations, buildings with greater energy efficiency are in demand, incentivizing building owners to invest in these improvements. The impacts of and benefits from mandatory disclosure laws may take some time to be realized. In some instances, the disclosure of energy performance data is not required from building owners until a full year's worth of data has been collected. Even after the related upgrades are implemented, time is needed to accurately verify the improved energy performance. Yet, even with these delays, markets respond, as building owners realize that the rules are out there or approaching, and tenants are more aware of their options.

In the United States, California was the first jurisdiction to launch a commercial benchmarking law, and several other jurisdictions, such as New York City, Seattle, and Washington D.C., have followed California's path. In Europe, Denmark was the first country to implement mandatory disclosure rules (in 1997), for all residential and commercial buildings. In the European Union (EU), the Energy Performance of Buildings Directive (EPBD) requires that office buildings 1,000 square meters or more track their energy performance and post an energy efficiency label. These labels vary from equivalent labels used in the United States, as they indicate the "as designed" energy efficiency of buildings, as opposed to the actual energy performance of the building. Each EU member country is utilizing its own methods and rules to adhere to these requirements. Effective disclosure programs in France and the United Kingdom are encouraging other European countries to follow suit. In Australia, buildings are becoming so efficient that it might be necessary in the near future to expand the current 1–5 rating category. Additionally, various areas in Canada and also

Asia are starting to look into practicing these regulations. It is expected that more and more countries, regions, and cities will follow suit [11].

2.2.1 Residential Buildings

The primary energy consumption of the residential sector plays a significant role within the overall buildings sector. In 2009, residential primary energy consumption was 54 percent of the consumption in the buildings sector. Residential energy demand is mainly driven by space heating (45 percent), water heating (18 percent), and space cooling (9 percent). Space heating and cooling combined account for 54 percent of site energy consumption and 43 percent of primary energy consumption [9].

The per square foot energy use of homes are improving with those built between 2000 and 2005 using 14 percent less energy than homes built in the 1980s, and 40 percent less than those built before 1950. Some of these efficiency improvements are being offset by the trend towards larger home sizes. Single-family homes built between 2000 and 2005 are on average 29 percent larger than those built in the 1980s, and 38 percent larger than those built before 1950. Moreover, there is a difference in the energy consumption profile of single-family and multi-family (apartments) homes. On average, multi-family homes have been seen to use 40 percent less energy than single-family homes, as they tend to have about half the floor space. However, even though multi-family homes typically use less energy *per household*, they use more energy *per square foot of heated space*. Households in the Midwestern and Northeastern parts of the United States have displayed the greatest energy consumption per household, as the average house size is larger in the Midwest, and as the Northeast has a larger share of older homes and smaller share of newer homes [9].

The U.S. Energy Information Administration's (EIA's) Residential Energy Consumption Survey (RECS) has identified several trends over past few decades relating to homes in the U.S. and their energy use. While the quantity of houses has steadily increased over time, the average energy consumption per housing unit reached an inflection point after which it started to decrease, indicating improved efficiencies. Over the past couple decades, both the number of households and their square footage has increased, while there has also been an increased reliance on the use of electronics. Even with all these burdens improvements in efficiency for space heating, air conditioning, and major appliances, and better insulation methods have offset the negative impacts and led to decreased energy consumption for households [12].

EIA's RECS shows that while homes built in the 1970s and 1980s were generally less than 1,800 square feet in size, by the 1990s average home size increased to 2,200 square feet and to 2,465 square feet by the 2000s. In addition, the ceiling height of new homes is also increasing—only 17 percent of homes built in the 1970s had higher ceilings (compared to conventional eight-foot ceilings), while 52 percent of new homes in the 2000s have higher ceilings. As the square footage of a house increases, there is an increased need for heating, cooling, and lighting, and use of various appliances. Moreover, it has been observed that parts of the nation which experience extreme winter weather conditions also tend to have larger homes, adding to the heating burden [12].

Even though most home appliances have become more efficient over the past three decades, the average U.S. household now uses more consumer electronics. The use of computers at home has shown a rapid increase. While not many households had computers in 1978, by 2009, 76 percent of U.S. homes had at least one computer, and 35 percent even had multiple computers. Similarly, in 1978, most households had only one television, but the average household in 2009 had 2.5 televisions [12].

Homeowners in the United States who engage in remodeling activities are also displaying energy conscious tendencies. It has been reported that in green home remodeling efforts, efficient HVAC systems and building envelope improvements (such as more efficient windows) are the most widespread elements to be used. Even though these improvements usually have higher initial costs, over the long run they provide cost savings from energy savings and improved comfort for the residents [10].

Globally, energy consumption in buildings is expected to rise from close to 32,000 TWh to about 51,000 TWh by 2050, with over half of this consumption in residential buildings. This trend pressures power companies to find ways to accommodate new demand, governments to consider carbon-related legislation, and households to incorporate the consequences of rising energy demand into their lives and their pocketbooks. Throughout the world, different factors such as climate considerations, residential building preferences, and diverse supply chains cause the markets to be very unique. However, the International Code Council has developed the International Energy Conservation Code (IECC), which sets minimum design energy performance standards for the building shell, space conditioning, lighting, and appliance operation for a variety of global climate zones, and has been adopted in a variety of territories globally. The market value of energy efficient homes globally through 2020 is expected to reach between $80 billion and $90 billion. While a rapid increase is seen as the overall global trend, the main contributors to this increase are expected to be the European Union and North America [13].

2.2.2 Commercial Buildings

In the commercial buildings sector, office space, retail space, and educational facilities represent approximately half of this sector's energy consumption, while the sector makes-up just under one fifth of the total national energy consumption. Space heating, lighting, and space cooling account for the majority of energy use in commercial buildings (27, 14, and 10 percent shares, respectively [9]).

Buildings used for medical services, food sales and food services typically have end-uses such as refrigeration and cooking, and tend to be occupied more hours per day and more days per week. Thus, while buildings used for these purposes represent only 8.5 percent of commercial floor space, they contribute to 19 percent of commercial primary energy consumption [9].

Lighting is the function that consumes the most energy in commercial buildings. Not only is there direct energy consumption through lighting, but the heat generated from the lighting fixtures creates an extra burden, increasing a building's cooling requirements This extra heat is a significant factor, as it can account for up to 42 percent of the cooling load of a typical commercial building. On the flip side, this extra heat helps with a building's winter heating requirements. It is not the most efficient pathway to achieving the needed heat, but is a factor nevertheless. The next largest factor in the energy use of commercial buildings are space heating. Water heating, ventilation, and office equipment use follow, with less of an energy demand. The energy demand of commercial buildings varies widely due to the variable nature of operations in these types of buildings. For example, water heating is a key function and energy consumer in hospitals and hotels, but not as much in offices or retail stores. The timeframe and duration of a commercial building's energy demand is also a factor. Some buildings, such as hospitals, operate on a 24-hour basis, so they have a continuous energy demand. On the other hand, buildings such as concert halls have a short, concentrated timeframe where energy demand peaks. Distinctions like these need to be fully understood in order to identify the opportunities for technology solutions and related practices to reduce energy demand in commercial buildings [7].

Demand for the services provided by energy to a building, over time, has both increased in scale and changed in structure. End-uses such as air conditioning, household appliances, office equipment, telecommunications devices, and entertainment units have moved from being novelties, to luxury items, to everyday items. This trend means that the demand for electricity and natural gas for these uses will continue to grow [7].

While government policies and incentives were the main motivation behind implementing building efficiency measures, this influence is decreasing over time, especially with commercial buildings. Owners of large corporate buildings are realizing the *business advantage* of efficiency enhancements and green building practices. While energy efficiency is only one of the components of a green building design, it is becoming one of the most important factors and driving the green building market, both in the commercial and residential sectors [10].

2.2.3 Global Building Efficiency Trends

The Institute for Building Efficiency, in partnership with the International Facility Management Association, the Urban Land Institute, and 16 strategic partners from various countries—conducted its yearly "Energy Efficiency Indicator (EEI)" survey. The 2012 survey [14] was conducted on close to 3,500 respondents throughout the Americas, Europe, Asia, and Australia. EEI 2012 showcases the inclinations, priorities, and practices of executives and building owners throughout the world, who deal with investment decisions in industrial, commercial and public-sector buildings. Insights gained from their financial evaluations, factors driving their actions, and barriers they confront have provided valuable information on global energy efficiency trends.

In the various regions surveyed in 2012, over sixty percent of respondents said their organizations were investing in energy efficiency, with the United States and Canada having the greatest number of organizations investing in energy efficiency. However, responses from these two countries also indicate the lowest investment in renewables, while Europe, Australia and Brazil invested about equally in energy efficiency and renewable energy projects. In addition, while approximately half of the respondents from most of the countries surveyed specified that they intended to increase their energy efficiency investments over the next year, the highest response was from China and India (81 percent for China and 74 percent for India). Private sector organizations (industrial and commercial) were more inclined than the public sector in increasing their investments in energy efficiency.

According to the EEI 2012 survey the main motivation behind implementing energy efficiency-related measures was cost savings. In the United States, Canada, Brazil, and Australia, rebates and other incentives provided by the governments and utilities ranked second, whereas in China, these incentives were ranked third. Other countries and regions had different priorities. While energy security ranked second in Europe and China, public image/branding ranked second in India.

Rebates and tax incentives were indicated as the most effective policies to encourage energy efficiency improvements for buildings. However, the developed regions (the United States, Canada, and Australia) relied on these incentives more. The developing regions (Brazil, China, and India) also felt that more firm building codes/equipment standards, as well as implementation of green appraisal standards would be as important in promoting improvements in building energy efficiencies.

The trends observed in EEI's yearly surveys have pointed to five main barriers to energy efficiency investments:

- Lack of awareness of opportunities for energy savings.
- Lack of technical expertise to design and complete projects.
- Lack of certainty that promised savings will be achieved.
- Inability of projects to meet the organization's financial payback criteria.
- Lack of available capital for investment in projects.

However, these barriers have different importance when it comes to the different regions around the globe. While lack of funding was seen as the greatest barrier in developed countries, developing regions face issues related to lack of awareness, technical expertise, and certainty of savings.

The yearly EEI surveys have also demonstrated that globally, interest in green buildings is continuing to grow, with increasing numbers of executives and building owners indicating that they plan to apply voluntary green building certifications for both new constructions and for existing buildings.

2.3 Efficient Vehicles

As seen from Figure 2.1, improvements in vehicle fuel economies do not necessarily translate into constant amounts of fuel savings. The same unit of increase in fuel economy will provide larger benefits for vehicles that are in the lower

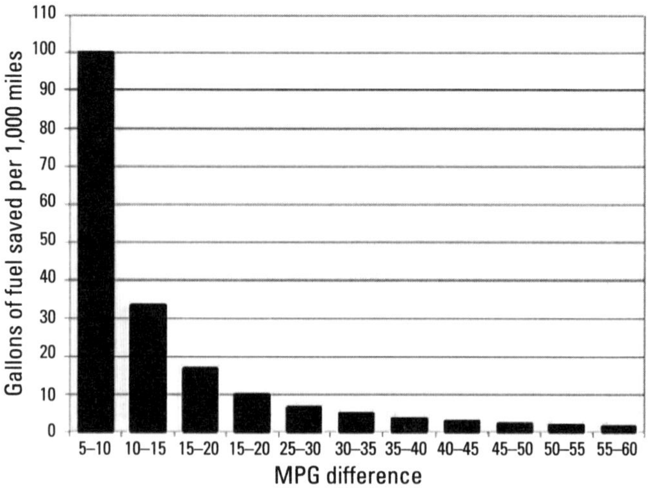

Figure 2.1 Vehicle fuel economy improvements vs. gallons of fuel saved [15].

range of fuel economies. Thus, as cars and their related fuel economies improve the impact of a unit of fuel economy improvement decreases [15].

Consumer surveys have demonstrated that the importance of vehicle fuel economy in consumers' purchasing decisions is once again starting to play a more important role. While vehicle fuel economy played an important role especially in the early 1980s, it declined thereafter. In 2011, vehicle fuel economy was chosen by 30 percent of the survey respondents as the most important characteristic when purchasing a vehicle, compared to 22 percent for vehicle dependability [15].

Over the past few decades, improvements observed in vehicle fuel economy in the United States have been smaller compared to improvements such as horsepower. While the horsepower of vehicles has increased dramatically (110 percent) from 1980 to 2010, fuel economy has only increased by 17.2 percent [15].

The general trend observed for the three U.S.-based auto manufacturers (Chrysler, Ford, and General Motors) is that vehicle fuel economies were below fleet averages, while horsepower was above fleet averages. In the five years spanning 2006 through 2010, Ford demonstrated the greatest advance in fuel economy (13 percent), while Chrysler and General Motors demonstrated only a 6 percent increase. On the other hand, Japanese auto manufacturers (Nissan, Toyota, and Honda) demonstrated fuel economies above fleet averages during the same period, and unlike the U.S. auto manufacturers, the Japanese have not increased horsepower (except Toyota). Other auto manufacturers (Hyundai in South Korea and Volkswagen in Germany) have displayed fuel economy improvement trends similar to the Japanese automakers [15].

A short-term but effective program aimed at improving the national vehicle fleet's fuel economy by taking inefficient vehicles off the road was the *Car Allowance Rebate System (also known as Cash for Clunkers)*, which was implemented between July 1, 2009 and August 24, 2009 in the United States. The program supplied federal rebate money to encourage consumers to trade-in their inefficient (EPA combined fuel economy of 18 miles per gallon or less) for new vehicles with better fuel economies. Vehicles purchased during this program had fuel economies that were on average about 59 percent above the vehicles that were traded-in. On a state-level basis, the average differences between the vehicles ranged from 7–8 miles per gallon to 10 miles per gallon (in California, where the price of fuel typically is higher than other parts of the nation) [15].

Another major enhancement related to vehicle fuel economy in the United States is not related specifically to technology, but to communications and outreach. In May 2011, the U.S. EPA and the National Highway Traffic Safety Administration (NHTSA) introduced a significant change to the fuel economy label, a degree of change not seen since the introduction of these labels 35 years ago. Two examples of these new labels are shown in Figure 2.2. As the variety of drivetrains and fuels used in today's cars has increased, these new labels have provided consumers more information and detailed comparisons (e.g., energy use and cost), to enable them to make more informed decisions. Some information in these new labels include: driving range and charging time for vehicles, how much fuel or electricity it takes to drive 100 miles, tailpipe emission ratings for criteria pollutants and greenhouse gases, and an estimate of fuel savings (or spending) over five years compared to an average new vehicle. The use of these new labels will become mandatory for the 2013 model year, but vehicle manufacturers may also chose to implement the labels starting in the 2012 model year [15].

The Corporate Average Fuel Economy (CAFÉ) is defined as the sales-weighted mean fuel economy of a manufacturer's fleet of new cars or light trucks in a certain model year (MY). This standard was initially approved by the U.S. Congress in 1975, to implement for cars beginning in MY 1978 and for light trucks in MY 1979. While overall the average of all cars and all light trucks has met or exceeded the standards each year, some manufacturers have not been able to reach the standards, while others have surpassed them. In December 2007, legislation was passed to raise the CAFÉ standards beginning in MY 2011 representing the first increase for cars since 1990. Auto manufacturers are provided another opportunity to meet these standards, by way of the Alternative Motor Fuels Act (AMFA) of 1988, which enables manufacturers to increase their calculated CAFÉ values by producing flex-fuel vehicles that can run on various alternative fuels. The act provides a manufacturer a credit of up to 1.2 miles per gallon toward its CAFÉ [15].

The average fleet-wide fuel economies required to meet the CAFÉ standards for cars, light trucks, and the combination of these two types of vehicles for MY 2012 through 2016 were set in May 2010. In November 2011, standards for MY 2017 through 2025 were proposed.[2] Proposed standards for cars in 2025 is 56 miles per gallon, while for light trucks it is 40.3 miles per gallon [15].

Moreover, the CAFÉ standards (starting in 2012), are to be based on a vehicle's footprint. The footprint of the vehicle is defined as the vehicle's track width times the wheelbase (distance between the wheels multiplied by the distance between the axles). The smaller the footprint, the stricter the standard, and vice versa. This method aims to help distribute the burden of compliance with the standard across all vehicles and manufacturers [15].

2. These standards apply to cars and pickup trucks less than 8,500 lbs. gross vehicle weight rating (GVWR), and sport utility vehicles and passenger vans less than 10,000 lbs. GVWR. GVWR is defined as the maximum operating weight of a vehicle that is specified by the manufacturer. It includes the vehicle's chassis, body, engine, engine fluids, fuel, accessories, driver, passengers and cargo but excludes trailers.

Figure 2.2 Examples of new vehicle fuel economy labels [15].

2.4 Efficient Industry

The industrial sector consumes a considerable amount of energy in the process of manufacturing a wide range of products. The aluminum, chemicals, forest product, glass, metal casting, mining, petroleum refining, and steel industries are among the most energy-intensive industries. Energy is lost throughout various stages in the manufacturing process, due to inefficiencies of the equipment, as well as mechanical and thermal limitations. Thus, optimizing the efficiency of these systems can result in significant energy and cost savings, and reduced carbon dioxide emissions. Even small improvements in efficiency can yield large energy savings and reduce industry's carbon footprint.

The industrial sector has large energy demands, thus providing an opportunity to meet good portions of that demand through energy efficiency, but this may be easier said than done, due to some related challenges. For example, industrial companies typically demand internal rates of return on capital investments as high as 20 percent [16], making the typically higher-cost initial investments for efficiency measures harder to implement. Moreover, as the industrial sector is global, some energy saving policies might prompt these industries to relocate to other countries.

When it comes to industrial facilities, there are several drivers in implementing energy efficiency measures. These facilities are businesses with many expenses and have the end goal of being profitable. Thus, energy is another cost (and a sizeable one) for them to try to reduce. At the same time, energy efficiency may be thought of as reducing wastefulness, which, if uncontrolled, negatively impacts the bottom line of the business. Especially with the highly variable costs of fuels used in producing the energy required to operate a facility, reducing energy consumption through efficiency measures provides industry protection against related risks and vulnerabilities. Managing energy consumption and associated costs provides the business a competitive edge, while also gaining the support of investors, and benefiting consumers.

The Energy Star Portfolio Manager tool has been an effective tool in tracking and managing building energy use. Between the years 2008 and 2011, an average annual savings of 2.4 percent and a total savings of 7 percent were attained with the use of the tool by 35,000 buildings in the United States. It has been verified by building energy benchmarking data analyses that buildings that benchmark their energy performance on a continuous basis do save energy. Analyses conducted by the EPA have demonstrated the extent of financial value achievable through even a few percentage points of energy savings. Analysis results have shown that a 2.4 percent savings for three successive years contributed to a cumulative energy cost savings of $120,000 for a 500,000 square foot office building, $2.5 million for a medium-box retailer with 500 stores, $4.1

million for a full-service hotel chain with 100 properties, or $140,000 for an 800,000 square foot school district [17].

The adoption of Energy Star appliances varies widely among different types of appliances. In 2011, Energy Star-labeled dehumidifiers and dishwashers achieved a market penetration of more than 95 percent, which was the highest rate of adoption, while water heaters, desktop computers and freezers had the lowest adoption in the marketplace [17].

2.5 Impacts

A term that frequently emerges when energy efficiency is under consideration is the "rebound effect." The term is used to describe a situation where benefits realized from energy efficiency measures unintentionally result in these benefits being reduced or even completely eliminated. For example, as vehicles get more efficient, they are able to run on less fuel. This in turn incentivizes the vehicle to be used *more*, as operating costs have now decreased, which consequently results in reducing the energy savings that had previously been attained. The same would hold true for efficiency improvements in homes, commercial buildings, industry, and consumer products.

In general, rebound effects can be categorized as "direct" and "indirect" [18]. Direct effects are realized from the use of an efficient product or the implementation of efficiency measures such as increased driving of an energy efficient car. An indirect effect is when consumers spend the money saved through energy efficiency measures, purchasing another energy-consuming product. Another indirect effect occurs when industrial processes become more efficient resulting in decreased production costs, increasing funds available for expansion of the industrial facility. In addition, as the production costs are lowered, demands for the products may increase.

Various studies have been conducted in an attempt to estimate the extent of direct and indirect rebound effects. Direct rebound effects were in general found to be 10 percent or less. There are more uncertainties involved with indirect effects, but the best estimate is around 11 percent. Thus, the two types of effects combined may add up to approximately 20 percent. However, even if this 20 percent loss is realized, the goal reached is still more advantageous than if the efficiency measures were not applied. There would still be an 80 percent improvement in efficiency, and the 20 percent would contribute to benefiting the consumers and the economy [18].

2.6 Market Status

Several programs, incentives, and stakeholders are having an impact on the energy efficiency market, which are highlighted in the following sections.

2.6.1 Leadership in Energy and Environmental Design (LEED)

Various green building certifications are used throughout the world to measure, verify and label a building's sustainability. In the United States, the Leadership in Energy and Environmental Design (LEED) Green Building Certification Program and the ENERGY STAR designation for homes and buildings are the most widely known and used programs [10]. LEED has long been the standard for sustainable building practices, and in the United States, the expansion of LEED-certified federal buildings has started to parallel the development of LEED-certified buildings in the private sector. In just the first eight months of 2012, the number of federal LEED-certified buildings increased about 51 percent, to 821, compared to the 544 federal buildings certified in all of 2011 [19].

Green building certification programs are spreading throughout the world, however, the U.S. Green Building Council's (USGBC's) LEED[3] is the only global certification system. While in some markets, LEED is being adapted to the local market circumstances, in many others the same scoring and rules used in the United States are being applied. For multinational companies operating in several different countries and markets, such consistency is a benefit. This standardization remains a challenge, as not all materials and services that are part of the U.S.-based LEED system are developed in other parts of the world. In an effort to overcome these challenges, the partnership-based *LEED International Program* was launched in 2010. In this program the USGBC provides support to the various national green building organizations that have signed on to the partnership in adapting LEED to local conditions [11].

While *LEED for New Construction (LEED-NC)* was initially the only LEED-related building rating available, the system has expanded to include improvements to existing buildings, such as core and shell improvements, renovations to commercial interiors, and enhancements to operations and maintenance practices. Half of the LEED certified commercial buildings in the United States are owned by for-profit organizations, while the rest are owned by entities such as state/local governments, the Federal Government, and nonprofits [9].

3. LEED does not only consider energy efficiency issues. It evaluates a building's design, construction and operation based on many environmental and energy factors, and promotes reduced environmental impact, improved occupant well-being, and increased profitability.

2.6.2 ENERGY STAR®

ENERGY STAR® is a program administered jointly by the U.S. Environmental Protection Agency (EPA) and the U.S. Department of Energy (DOE), initiated in 1992 as a voluntary labeling program, with the goal of distinguishing and thus promoting energy-efficient products to help reduce greenhouse gas (GHG) emissions. While initially only computers and monitors were labeled under this program, it gradually expanded to covering additional office equipment and residential heating/cooling equipment. Currently, the ENERGY STAR® label covers a wide range of products, including major appliances, home electronics, lighting, and even homes and commercial/industrial buildings. In 2011 alone, the emission of 210 million metric tons of GHGs have been eliminated through this program (compared to 0.7 million metric tons in 1993). This is the equivalent of removing the annual emissions of 41 million vehicles, while also saving consumers $23 billion on their utility bills [20].

2.6.2.1 ENERGY STAR® for Buildings

The ENERGY STAR® rating system started out as an approach to evaluate appliances and electronic devices based on their energy efficiency. However, it has evolved into a tool for also evaluating the energy efficiency of buildings. The energy performance of commercial buildings can be compared to similar types of buildings and rated on a scale of 1 to 100. ENERGY STAR® labeling is given to buildings that achieve a score of 75 or higher. Similarly, single-family homes may be evaluated under the ENERGY STAR® program, where homes that earn the label are specified by EPA as "at least 15% more energy-efficient than homes built to the 2004 International Residential Code (IRC), and also include additional energy-saving features that typically make them 20%–30% more efficient than standard homes" [10].

The American Council for an Energy Efficient Economy (ACEEE) published their International Energy Efficiency Scorecard 2012 [21], in which they analyzed the 12 largest economies around the globe and used 27 metrics to assess how efficiently these countries use energy. The countries analyzed collectively represent more than 78 percent of global gross domestic product (GDP), 63 percent of global energy consumption, and 62 percent of the global carbon dioxide-equivalent emissions. The metrics used were a mix of performance metrics that gave *quantifiable* results, and policy metrics that represented the best practice policies. The performance metrics included those such as overall country energy use per that country's GDP, residential building energy use per square foot, and average mileage (miles per gallon) of passenger vehicles. The policy metrics included factors such as the presence and implementation of a national energy savings target, vehicle energy efficiency standards for appliances, and vehicle fuel economy targets. The metrics were divided into three main sectors that represent the main energy consumption areas; buildings, industry and transportation. As some metrics are more cross-cutting, in addition to the three main sectors, a "national efforts" category was also developed and considered. The maximum possible score for each country was determined as 100, and points were allocated across the four categories under consideration, after which each country received a total score and was ranked.

As expected, some countries outperformed others, but more importantly, results of the ACEEE Scorecard highlighted that each country still had significant opportunities for further improvements. Even though several countries received full scores in some metrics, the average score was only 54 points out of a 100. Germany received the highest score for the "national efforts" category (19 out of 25 points). China received the highest score for the buildings category (23 out of 28 points). The United Kingdom received the highest score for the industry category (18 out of 24 points). The transportation category was a tie between several countries (each with 14 out of 24 points)—Italy, China, Germany, and the United Kingdom. The United Kingdom had the highest score overall, with 67 out of 100. The United States was lower in the rankings (with an overall score of 47 out of 100), ending-up as ninth out of the 12 countries, mainly due to its lower score in the transportation category (even though it scored relatively well in the buildings and industry sectors). Over the last decade, the United States has demonstrated progress towards better energy efficiency, in areas such as building codes, appliance standards, government-industry partnerships, and the more recent vehicle fuel economy standards. However, as the ACEEE Scorecard has highlighted, progress in energy efficiency in the United States is falling behind that of other countries such as Germany, Japan, and China. Moreover, progress some countries demonstrate in energy efficiency enables them to have a competitive edge over other countries. By being more efficient, these countries use fewer resources to achieve the same end results, thus decreasing costs while saving natural resources. Being on the lower end of the efficiency spectrum means that more resources are wasted and costs are increased.

In 2010, close to a quarter of all single-family homes permitted for ENERGY STAR® certification were qualified for the label. In Hawaii, Nevada, Iowa, and Arizona more than half of all new homes qualified for this label. At the same time, the number of homes retrofitted under the Home Performance with ENERGY STAR® program demonstrated rapid growth resulting in a 41 percent increase from 2009 and a 158 percent increase from 2008. In the commercial buildings sector, the ENERGY STAR® label was awarded to buildings that represented 3.7 percent of all commercial floor space in the United States. In terms of ENERGY STAR® label qualified commercial building floor space, office buildings and K–12 schools are the largest contributors [9].

2.6.2.2 ENERGY STAR® for Homes and Residential Buildings

The ENERGY STAR® rating and labeling system is now also being applied to homes and the EPA has been setting more demanding requirements for qualified homes, positioning them to perform up to 30 percent better than typical new homes. In addition, multifamily high-rise buildings that are new or substantially renovated became eligible to earn the ENERGY STAR® label in 2011, with requirements designed to be at least 15 percent more efficient than the building energy code.

The comprehensive home evaluation and retrofit program, Home Performance with ENERGYSTAR® (HPwES), has reached its 10th year of implementation in 2012, with 200,000 home improvement projects completed in that timeframe [22]. These home improvement projects start with a detailed energy audit of the home, performed by a certified HPwES contractor. This contractor carries-out several diagnostic tests and compiles findings in an energy audit specific to the home and includes recommendations for energy-saving home retrofits, as well as related energy savings. The home owner may then coordinate with the contractor on the most suitable home improvement options.

2.6.2.3 ENERGY STAR® for Consumer Products

The products that qualify for an ENERGY STAR® labeling provide energy savings of 65 percent, relative to their conventional counterparts. The labeling system and its benefits have become widely recognized by the American public. Over 80 percent of American consumers now recognize the label, and about 75 percent indicate that the label is an important factor in their decision. Qualified products purchased since 2000 have shown rapid growth, reaching over 4 billion items purchased in 2011 compared to the less than half billion items purchased in 2000. EPA has also implemented rigorous testing and certification requirements (implemented by 400 laboratories from around the world and 21 certification bodies) aimed at ENERGY STAR® qualified products for over 60 product categories [20].

2.6.2.4 ENERGY STAR® for Businesses

In 2011, over 7,500 commercial buildings and plants received the ENERGY-STAR® label. These certified buildings typically use 35 percent less energy and account for 35 percent fewer GHG emissions than average buildings [20]. The EPA also has an interactive energy management tool for businesses, the ENERGYSTAR® Portfolio Manager™, which allows the online tracking and management of energy and water consumption associated with a portfolio of buildings. This tool can help determine buildings that are not performing to standards, verify improvements in efficiency, and help guide related investments.

The EPA has revealed substantial energy performance improvements in energy efficiency across the cement industry, including a 13 percent improvement in energy intensity. Additionally, many industrial plants in various sectors have committed to the *ENERGYSTAR® Challenge for Industry*, a national call to industry to improve their energy efficiency by 10 percent within five years [23]. In 2011, 60 plants met or exceeded this target, saving 16 trillion Btu in energy and preventing the equivalent of nearly 1 million metric tons of GHG emissions [20].

2.6.3 Energy Service Companies (ESCOs)

Efficiency-related projects in the buildings sector also benefit from the Energy Service Company (ESCO) industry, as ESCOs aid with the long-term financing that energy efficiency-related projects typically require. The majority of business for ESCOs (over 70 percent) is in the energy efficiency sector [11].

The ESCO business has been expanding in the United States since the 1970s, and has also been adopted in some Asia Pacific and Southeast Asian countries. The adoption and growth in Asia is expected to continue, due to business interest in the low-risk approach to energy efficiency, along withgovernment support. The emergence of the ESCO business has mainly been based on its reliance on the public sector. In the United States, the federal building market is the source of approximately one fourth of the ESCOs' business. Public housing is also expected to contribute to the ESCO business, as the U.S. Department of Housing and Urban Development has an initiative to build partnerships with ESCOs and to provide support to local housing authorities [11].

2.7 Challenges and Potential Solutions

The basic premise of increasing end-use efficiency, and thus reducing energy usage, conflicts with utilities' sales and revenue-based interests. Thus, appropriate policies are needed to help overcome this situation. With the right policies, positive investment environments may be enabled for utilities, funding may be authorized for energy efficiency efforts, while ensuring that building energy codes and appliance efficiency standards are employed. Different rate-making structures, accompanied by strong incentives for the utilities are typically used to balance the conflict between the implementation of energy efficiency measures and utilities' basic sales needs [1].

In order for energy efficiency measures to be implemented successfully, several market barriers that influence consumer choice and behavior need to be overcome [1]:

- *Awareness:* Consumers may not be aware of or educated about different energy management options and potential savings to be realized by implementing energy efficiency measures.

- *Availability:* Efficient products may not being produced at all, not produced in quantities adequate for the demanding market, or not effectively marketed. There may also be instances where incentive programs are detached from the value proposition to the end-user (such as when the decision-maker and the benefiting party are not the same.) For example, a home builder may be more concerned about building a home for the lowest cost, and does not have an extra incentive (or benefit) to also implement energy efficiency-related measures, as generating a profit on investment is the main priority.
- *Accessibility:* This barrier is slightly different than the availability barrier, in that while an energy efficient product might be produced and available, there might be stocking issues, or products might not be displayed at retailers in areas where they can be easily found by consumers. Or, there may be inadequacies in support mechanisms such as contractors with the required skills and experience levels to install and sell these energy efficiency technologies and solutions.
- *Affordability:* Energy efficient technologies are typically more expensive than the conventional technologies, and high initial costs are a major barrier in consumers' eyes, even when a viable portfolio of incentives is presented by efficiency program administrators.

2.7.1 Energy Management through ISO 50001

The International Organization for Standardization (ISO) is a group of national standards bodies from 160+ countries, with a range of over 18,000 standards that cover the economic, environmental, and social aspects of sustainable development for all sectors. Many experts are involved in setting ISO standards meaning that global consensus on the state of the art in the subject of that standard has been reached [24].

A recent standard among the portfolio of ISO standards is *ISO 50001: Energy Management Systems.* ISO 50001 is a voluntary international standard developed by the ISO and provides requirements for energy management systems. The standard is applicable to organizations of all sizes in the private and public sectors regardless of geographical location, with the potential to impact up to 60 percent of global energy use. While organizations do not necessarily have direct control over factors such as energy prices, global economy, or government policies, they do have control over how they manage energy in their facilities and processes [24].

ISO 50001 intends to assist organizations with [24]:

- Making better use of their existing energy-consuming assets;

- Creating transparency and facilitating communication on the management of energy resources;
- Promoting energy management best practices and reinforcing good energy management behaviors;
- Assisting facilities in evaluating and prioritizing the implementation of new energy-efficient technologies;
- Providing a framework for promoting energy efficiency throughout the supply chain;
- Facilitating energy management improvements for greenhouse gas emission reduction projects; and
- Allowing integration with other organizational management systems such as environmental, and health and safety.

ISO 50001 is based on the "plan-do-check-act" continual improvement protocol that is followed by all ISO standards, and can be described as follows [24]:

- *PLAN:* Conduct the energy review and establish the baseline, energy performance indicators, objectives, targets and actions necessary to deliver results in accordance with opportunities to improve energy performance and the organization's energy policy.
- *DO:* Implement the energy management action plans.
- *CHECK:* Monitor and measure processes and the key characteristics of operations that determine energy performance against the energy policy and objectives, and report the results.
- *ACT:* Take actions to continually improve energy performance and the energy management system.

2.8 Emerging Opportunities

Better methods of managing buildings and improving building efficiencies through the use of technologies that utilize wasted heat are emerging as valuable opportunities in the energy efficiency arena

2.8.1 Building Energy Management Systems (BEMS)

An valuable emerging tool to evaluate and improve the efficiencies of buildings is a Building Energy Management System (BEMS). A BEMS is a computer-based system that automates and manages a building through various monitoring devices such as switches and sensors. While traditional building management systems adjust temperatures in the building according to a preset schedule, a BEMS utilizes more advanced sensors and analytics to more accurately control the building and maximize its energy efficiency. Further, the implementation of a BEMS typically will cost less than applying other conventional efficiency measures, while minimally disrupting the building's occupants. These factors are especially appealing to commercial buildings such as schools and hospitals, where funds are tight and occupant stability and safety are important. Bigger and older buildings are prime candidates for employing a BEMS. In bigger facilities, the fixed cost of the control software may be distributed over the term of the project, and in older buildings, there is a greater untapped potential to improve efficiency.

2.8.2 Combined Heat and Power (CHP)

Combined heat and power (CHP) is a pathway that may be applied to a range of technologies, providing increased efficiency by generating electric power and useful thermal energy from a single fuel source. As an alternative to purchasing electricity from the electric grid and burning fuel in an on-site boiler to produce thermal energy, CHP may be used by an industrial or commercial facility to deliver both the electricity and heat needs of the facility. Figure 2.3 displays the efficiency benefits of a typical CHP system. CHP applications may reach efficiencies in the range of 65 percent to 75 percent, a considerable enhancement to the average of 45 percent if electricity and heat were to be separately provided. The average efficiency of power generation in the United States has not improved much since the 1960s, due to the large amount of energy lost as wasted heat from power generation. In fact, the amount of this wasted heat is greater than the total energy use of Japan. Thus, CHP offers an opportunity to

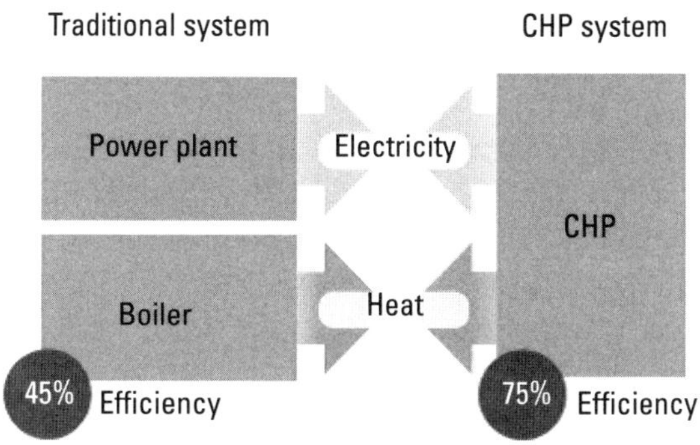

Figure 2.3 Efficiency comparison of traditional system vs. CHP system [25].

capture and make use of this wasted energy, improving overall efficiency, saving money, and havingless of a negative impacton the environment [25].

Even though the CHP has been in use for a while and continues to expand in the United States, it still is an underutilized resource, representing approximately 8 percent of U.S. generating capacity compared to over 30 percent in countries, such as Denmark, Finland, and the Netherlands. Policymakers at the federal and state level are now beginning to recognize the potential benefits of CHP and the important role it could play. Business is also beginning to consider CHP as a fruitful investment, especially if they are facing sizeable costs to upgrade outdated coal and oil-fired boilers. Additionally, the economics of CHP are improving as a result of the changing outlook in the long-term supply and price of North American natural gas, which is a fuel preferred for many CHP applications [25].

A federal-level policy initiative set forth by the Obama Administration in 2012 calls for reaching 40 GW of installed CHP by 2020. Reaching this goal would have several significant implications for the United States [25]:

- Total CHP capacity in the United States would be increased by 50 percent in less than a decade;
- Consumers would save $10 billion a year compared to current energy use;
- One quadrillion Btus (Quads) of energy would be saved;
- Emissions would be reduced by 150 million metric tons of CO_2 annually, which would be the equivalent of the emissions from over 25 million cars; and
- $40–$80 billion in new capital investment in manufacturing and other facilities would be created over the next decade.

The CHP pathway has the advantage of being a distributed energy production pathway, being strategically located at or near the point of energy use. Of the existing CHP capacity in the United States, 87 percent is located at industrial facilities. This means that there is a yet to be tapped potential for the deployment of CHP at commercial or institutional facilities, such as schools, hospitals, district energy systems, and military installations. Providing energy needs onsite prevents the losses related to the transmission and distribution of electricity purchased from the grid, while postponing or even eliminating the need for new investment in the grid infrastructure. The higher overall efficiencies provided by CHP along with the elimination of transmission and distribution losses translates into reduced primary energy use and lower GHG and criteria pollutant emissions (especially if lower-carbon fuels such as natural gas are used). It has been demonstrated that CHP is capable of providing energy and carbon savings equivalent to comparably sized solar photovoltaic, wind, and natural gas combined cycle (NGCC) systems, at a capital cost that is lower than solar and wind and on par with NGCC. CHP may also provide lower energy costs for the user by displacing higher-priced purchased electricity and boiler fuel with lower-cost self-generated power and recovered thermal energy, if the savings achieved from power and fuel costs are more than the added capital, fuel and other operating and maintenance costs of the CHP system [25].

Natural gas is the most common fuel used in CHP systems in the United States, acting as a fuel for 72 percent of installed CHP capacity. However, other resources such as biomass, process wastes and coal are also utilized. When compared to average electricity generated from fossil resources, the entire existing base of CHP saves 1.8 Quads of energy and eliminates 240 million metric tons of CO_2 emissions each year (equivalent to the emissions of over 40 million cars) [25].

Distributed CHP systems at industrial and municipal sites were common in the early electric power industry in the United States. The power industry then began to build larger central station facilities to take advantage of increasing economies of scale, restricting the use of CHP mainly to a few industries (paper, chemicals, refining and steel) which had high and relatively constant steam and electric demands and access to low-cost fuels. In this business climate, utilities did not have much incentive to encourage onsite generation by the customer. In 1978, during the era of the oil crisis, the Public Utilities Regulatory Policies Act (PURPA) was passed with a goal of stimulating greater energy efficiency, by supporting energy efficient CHP and small power production from renewable. PURPA required electric utilities to interconnect with "qualified facilities" (QFs) and to purchase excess electricity from these QFs at the "avoided cost" of the utilities. In addition, Congress also made tax credits available for investments in cogeneration equipment, through the Energy Tax Act of 1978 and the Crude Oil Windfall Profits Act of 1980. These initiatives expanded the use of CHP, resulting in an increase in installed capacity from approximately 12,000 MW in 1980 to over 66,000 MW in 2000. In the early 2000s, with the restructuring, or deregulation, of wholesale electricity markets in some parts of the nation, independent power producers were allowed to sell directly to the market without the need for QF status. In specific states this trend resulted in market uncertainty, delaying energy investments and thus slowing CHP development. At the same time, natural gas prices were rising and becoming increasingly volatile, further inhibiting the market for CHP development, with investment in new CHP capacity slowing drastically in the 2004/2005 timeframe, and to this day [25].

Evaluations have shown that there is about 65 GW of technical potential for additional CHP at existing industrial facilities, and over 65 GW for commercial and institutional facilities, resulting in a total technical potential of approximately 130 GW. In fact, the technical potential in industrial facilities could double to 130 GW if systems are sized to the thermal demand without a cap in power output, and excess electricity generated is exported to the grid or sold to adjacent users. Figure 2.4 displays the technical potential for additional CHP at a variety of types of industrial and commercial facilities [25].

In the United States, several factors have been affecting the electricity production market, in turn affecting CHP deployment. Some of the key factors are [25]:

- Natural gas prices are on a significant downward trend (due to shale gas discoveries) and are expected to remain low for several years. The shale gas market in the United States has multiplied by a factor of 14 since 2005, displacing imports and offsetting declines in other North American production resources.
- Coal is displaying moderately rising prices.
- Projections for electricity demand growth have been reduced.
- Coal-based power plants are starting to retire.

The EPA has finalized power sector air regulations, requiring investments in pollution control technology at fossil-fired plants that currently lack modern controls.

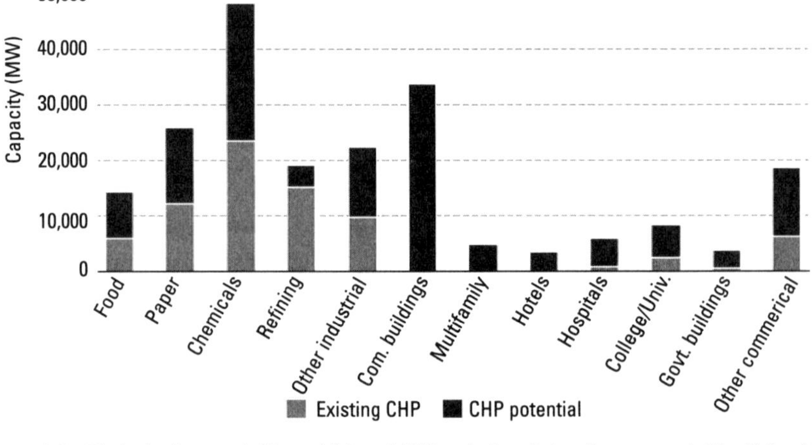

Figure 2.4 Technical potential for additional CHP at industrial and commercial facilities [25].

Especially over the past two years, several coal-based power plant owners in the United States have announced plant retirements. Facilities will inevitably need to invest capital to improve or replace aging boilers, in order to comply with pollution standards or to tackle aging capital equipment. This opens-up a window of opportunity for the deployment of CHP, as it is often a better investment, being cleaner and more energy efficient [25].

There are several remaining technical and regulatory barriers to the further advancement of CHP in the United States [25]:

- *Unclear Utility Value Proposition:* Even though facilities that install CHP gain considerable flexibility through their self-generation capacity, they remain reliant on the electric grid and the utilities for backup power needs. Thus, utility policies, rate structures, and actions can have an important impact on a CHP project's economics. In addition, interconnection processes can delay the project development process and add expenses.
- *Limited CHP Supply Infrastructure:* As CHP is not currently a major consideration for most energy developers and equipment suppliers, there is a limited supply infrastructure. If deployment is increased, costs will come down, helping expand this infrastructure.
- *Market and Nonmarket Uncertainties:* Various uncertainties impact the economics of CHP projects, including fuel and electricity prices, regional/national economic conditions, market sector growth, utility and power market regulation, and environmental policies. Moreover, CHP may actually increase onsite emissions while reducing regional emissions.
- *End-User Awareness and Economic Decision-Making:* CHP systems typically are not the main business focus of the end user. Thus, they are sometimes subject to higher investment hurdle rates than competing options. The full benefits provided by CHP are not always properly realized, while decision makers may be wary of perceived CHP investment risks.
- *Local Permitting and Siting Issues:* CHP technologies are faced with various local zoning, environmental, health and safety requirements at the local site, including rules on air and water quality, fire prevention, fuel storage, hazardous waste disposal, worker safety, and building construction standards. This situation is further complicated by the fact that adhering to these requirements involves interacting with various local agencies including fire districts, air districts, and water districts and

planning commissions, who typically are unfamiliar with the technologies and systems.

References

[1] McLean-Conner, P., *Energy Efficiency: Principles and Practices,* Tulsa, OK: PennWell Corporation, 2009.

[2] U.S. Department of Energy-Weatherization and Intergovernmental Program, Accelerating Adoption of Energy Efficiency and Renewable Energy, DOE/GO-102010-3059, June 2010.

[3] U.S. Department of Energy-Office of Inspector General (Office of Audit Services), Audit Report: The Department of Energy's Implementation of the Energy Efficiency and Conservation Block Grant Program under the Recovery and Reinvestment Act: A Status Report, OAS-RA-10-16, August 2010.

[4] Website: U.S. Department of Energy-Weatherization and Intergovernmental Program, Energy Efficiency and Conservation Block Grant Program, Available at: http://www1.eere.energy.gov/wip/eecbg.html

[5] U.S. Department of Energy-Advanced Manufacturing Office, *Guide for Better Buildings, Better Plants Program Partners,* November 2011.

[6] Mackres, E., et al., "The Role of Local Governments and Community Organizations as Energy Efficiency Implementation Partners: Case Studies and a Review of Trends," © American Council for an Energy-Efficient Economy, February 2012.

[7] U.S. Department of Energy, Building Technologies Program: Planned Program Activities for 2008-2012, 2008.

[8] Randolph, J., and G. M. Masters, *Energy for Sustainability: Technology, Planning, Policy,* Washington, D.C.: Island Press, 2008.

[9] D&R International Ltd., *2011 Buildings Energy Data Book,* March 2012.

[10] McGraw Construction, *Energy Efficiency Trends in Residential and Commercial Buildings,* August 2010.

[11] Wapner, M., and C. Wheelock, *Building Efficiency: Ten Trends to Watch in 2011 and Beyond (Research Report, 4Q 2010),* Pike Research, 2010.

[12] Website: U.S. Energy Information Administration, Residential Energy Consumption Survey, Available at: http://www.eia.gov/consumption/residential/.

[13] Gibson, B. E., and K-A. Adamson, *Energy Efficient Homes (Research Report, 2Q 2012),* Pike Research, 2012.

[14] Institute for Building Efficiency, *2012 Energy Efficiency Indicator: Global Results,* June 2012.

[15] Davis, S. C., R. G. Boundy, and S. W. Diegel, *2011 Vehicle Technologies Market Report, ORNL/TM-2012/016,* Oak Ridge, TN: Oak Ridge National Laboratory, February 2012.

[16] Klustner, K., *Energy Efficiency: The Future is Now,* Kevin Klustner, 2008.

[17] *Energy Manager Today,* "Buildings Using Energy Star Portfolio Manager Realized Total Savings of 7%," October 19, 2012.

[18] Nadel, S., "The Rebound Effect: Large or Small?" © American Council for an Energy-Efficient Economy, 2012.

[19] *Environmental Leader: Environmental & Energy Management News,* "LEED Certifications for Federal Buildings Rise More Than 50%," October 19, 2012.

[20] U.S. Environmental Protection Agency, *ENERGY STAR® Overview of 2011 Achievements,* 2011.

[21] Hayes, S., R. Young, and M. Sciortino, "The © ACEEE 2012 International Energy Efficiency Scorecard," Washington, D.C.: © American Council for an Energy-Efficient Economy, 2012.

[22] *Home Performance with ENERGY STAR® Newsletter,* "Home Performance with ENERGY STAR® Celebrates a Decade of Success and Completes over 200,000 Projects," Summer 2012.

[23] Website: ENERGY STAR®, ENERGY STAR® Challenge for Industry, Available at: http://www.energystar.gov/index.cfm?c=industry_challenge.industry_challenge.

[24] International Organization for Standardization, *Win the Energy Challenge with ISO 50001,* 2011.

[25] U.S. Department of Energy and U.S. Environmental Protection Agency, *Combined Heat and Power: A Clean Energy Solution,* August 2012.

3

Renewable Electricity

3.1 Overview of Renewable Electricity

The economics of renewable technologies are (with the exception of biomass) substantially different from fossil-based technologies since renewables generally have high initial investment costs, modest O&M costs, and very low or no fuel costs. Conventional fossil-based facilities have lower initial investment costs and high (and in some cases even dominant) fuel costs which are very sensitive to price volatilities. Renewable technologies are also more sensitive to changes in the cost of capital and in financing conditions.

The Clean Energy Patent Growth Index (CEPGI) tracks the U.S. clean energy-related patents awarded on a quarterly basis since 2002. The second quarter 2012 evaluations [1] have highlighted the continued growth in the issuance of patents. In the United States, among various clean energy technologies, fuel cells continue to be in the lead (264 patents) with a substantial margin, although this gap has been narrowing over the past year. Solar technology patents closely followed fuel cells' lead, with 211 patents, while wind technologies were right behind solar, with 187 patents. Other renewable electricity-related technologies displayed a growing trend, but with a lower total number of patents issued (50 or less) [1].

Figures 3.1 through 3.3 provide a snapshot of trends related to all renewables, both in the United States and globally.

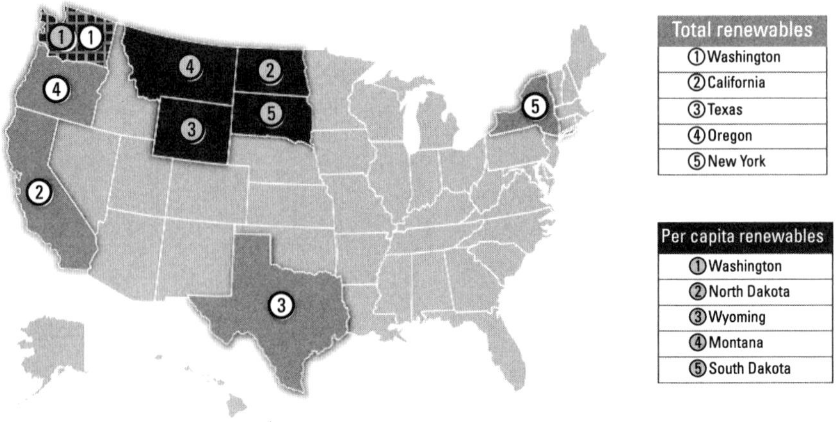

Figure 3.1 Top states for renewable electricity installed capacity (all renewables; 2011) [2].

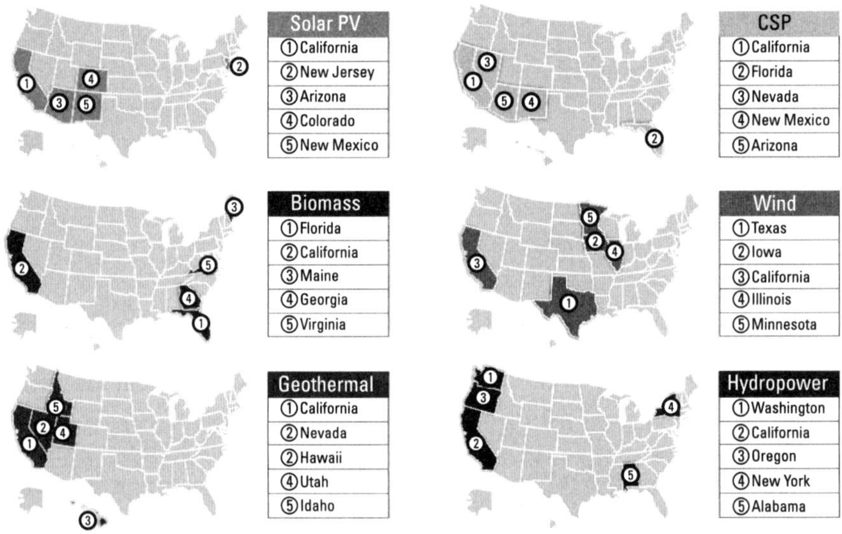

Figure 3.2 Top states for renewable electricity installed capacity (by technology; 2011) [2].

3.2 Solar Energy

Solar energy represents a vast resource, with the amount of energy supplied by the sun in one day being large enough to meet all the energy needs of the global population [3]. Approximately 885 million TWh worth of solar radiation reaches Earth's surface each year [4]. The technical potential of solar energy in most regions of the world is generally many times greater than current total

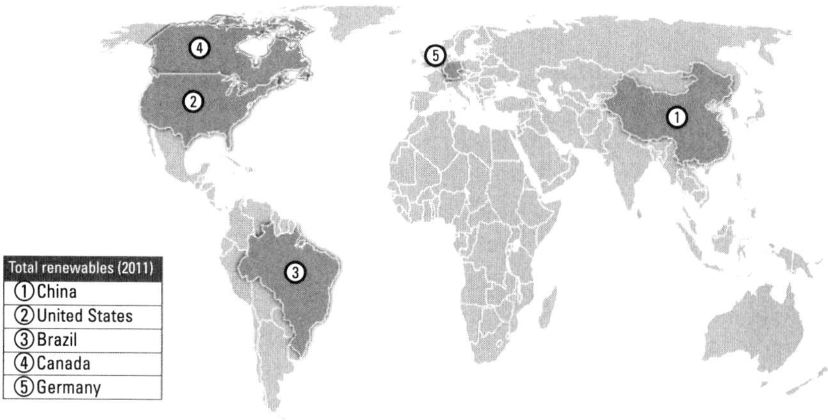

Figure 3.3 Top countries for renewable electricity installed capacity (all renewables; 2011) [2].

primary energy consumption in those regions. However, despite the vast resource potential provided by the sun and the recent growth of solar technology on the market, the contribution of solar energy to the global energy supply mix is still very small [5].

3.2.1 Basics

Solar energy may be utilized through passive or active methods. Passive solar involves collecting energy from the sun without converting its heat or light into other forms, by optimizing building design to make better use of daylight or heat. Active solar involves using photovoltaic (PV) technologies to convert the sun's energy into electricity, using solar thermal technologies to convert the sun's heat into electricity, or to use the heat directly [5]. PV is a technology that was invented in the 1950s and first used to power satellites. As the cost decreased from the initial high levels, PV systems were installed in off-grid applications. In the past decade, and especially in the last several years, grid-connected deployments have taken over the market and displayed rapid growth [6]. CSP plants were initially deployed in the United States in the early 1980s, and installations continued through the early 1990s. Many of these systems are still in operation today, but until recently, not many new systems had been deployed. Announcements for a several new plants projected to be completed between 2012 and 2015 have recently been made [6].

3.2.1.1 Resource Base and Characteristics

All of the energy that reserves of coal, oil, and natural gas hold can be matched by energy from just 20 days of sunshine. The energy provided by the sun before it reaches the earth's atmosphere is 1,300 Watts/m^2. From this point on, some of the sun's rays are reflected to space, while some are absorbed by atmosphere. By the time the sun's rays reach the earth's surface, its energy has been reduced to 1,000 Watts/m^2 (on a clear day at noon). When this value is averaged over the entire surface of the earth, for 24 hours per day for a year, each square meter of surface area accumulates approximately the energy equivalent of almost a barrel of oil each year, or 4.2 kWh/day of energy. This maximum quantity cannot be fully utilized, as it is impacted by location, cloud cover, seasons, and efficiencies of the solar technologies utilized, but this still represents a significant resource that can be used to meet energy needs [7].

3.2.1.2 Solar Photovoltaic (PV)

A photovoltaic (PV) system is, at the core, composed of solar cells. The combination of several of these cells forms a module, which is the smallest PV component sold commercially, and ranges in power output from about 10–300 Watts [8]. Several modules may also be combined, for a larger total power output, to form an array. A full PV system that is tied to the electric grid consists of one or more PV modules, an inverter which converts the system's direct-current (DC) electricity to the alternating current (AC) electricity that the grid is based on, and batteries to provide energy storage or back-up power. PV-based electricity may be variable, but it also corresponds well with peak electricity demand that is driven by cooling in summer and year round in hot countries. The horizontal irradiance reaching PV systems may be increased by either using tilting collectors (can increase irradiance by up to 35 percent, especially for latitudes lower than 30°S and higher than 30°N), or systems that track the sun for optimal resource capture. While both these methods can increase yield, they also add considerable additional expense [4]. At the moment, two main types of PV technologies are available in the market: crystalline silicon-based PV cells and thin film technologies based on a range of different semiconductor materials (amorphous silicon, cadmium-telluride and copper indium gallium diselenide) [5].

In a PV cell, the absorbed sunlight is transferred to electrons in the atoms of the semiconducting materials that make up the PV cell. The electrons are activated by this energy, and flow from where they are located to become a part of the electrical flow (current) in an electrical circuit. A built-in electrical field in the PV structure, called the p/n junction, provides the voltage required to drive this current through an external load (e.g., a light bulb). The p/n junction is formed by layering two different types of semiconductor material together. The n-type material has an abundance of electrons (thus a negative electrical charge), while the p-type material has an abundance of "holes" (thus a positive electrical charge). The combination of these materials causes the excess electrons to move from the n-type side to the p-type side [3].

PV systems which utilize the wafer-based crystalline silicon (c-Si) technology (either single crystalline or multicrystalline) are fully commercial. These solar cells dominate the market with their low costs, high efficiency, and wide range of well-established manufacturers. Silicon is an abundant element in the earth's crust and is a semiconductor material suitable for PV applications. Crystalline cells are in mass production and may potentially be produced at the rate of several hundred megawatts (MWs) or even gigwatts (GWs) per year.

PV systems based on thin film technologies are in the early market adoption phase. They generally include three main families: (1) amorphous (a-Si) and microcrystalline silicon (μc-Si); (2) Cadmium-Telluride (CdTe); and (3) Copper-Indium-Selenide (CIS) and Copper-Indium-Gallium-Diselenide (CIGS). The basic structure of thin film solar cells involves successive thin layers of solar cells deposited onto a large, inexpensive substrate such as glass, polymer, or metal. The use of the substrate allows them to use up to 99 percent less semiconductor material than crystalline cells to absorb an equivalent amount of sunlight, providing lower costs. However, the lower capital costs due to lower production and materials costs are offset to some extent by lower efficiencies. Thin film cells may also be packaged into flexible and lightweight structures, which may easily be integrated into building components, allowing for building-integrated PV.

Amorphous silicon and CdTe PV cells are the most developed thin film solar cells. A version of amorphous silicon solar cells is the multijunction thin-film silicon, which consists of a-Si cell with additional layers of a-Si and microcrystalline silicon applied onto the substrate. The μc-Si layer absorbs more light from the red and near infrared part of the light spectrum, thus increasing the efficiency by up to 10 percent. Cadmium Telluride thin-film PV solar cells have low production costs and efficiencies as high as 16.7 percent, thus making them the most economical thin-film technology currently available. CIS and CIGS PV cells offer the highest efficiencies of all thin-film PV technologies, with current module efficiencies in the range of 7–16 percent.

3.2.1.3 Concentrating Solar Power (CSP)

The main concept behind concentrating solar power (CSP) technology is the use of mirrors or lenses to concentrate the incoming rays of sun, which in turn heats a fluid and produces steam. The rest of the process is similar to conventional power plants, as the steam drives a turbine and generates power. Future iterations of the technology may not necessarily use this conventional steam cycle. While the steam cycle is similar to conventional power plants, the difference (and benefit) of CSP plants is that they utilize clean renewable energy from the sun to drive these reactions, rather than fossil fuels. CSP plants also generally come equipped with their own heat storage system, which increases the facility's capacity, while enabling in-demand electricity and better integration of renewable solar power with the grid. This could enable the technology to become competitors for conventional base- or intermediate-load power plants [9]. Besides providing renewable power, CSP is also capable of generating considerable amounts of valuable high-temperature heat for industrial processes, especially to help meet growing demand for water desalination in arid regions. CSP also holds potential for other diverse applications such as household cooking, small-scale manufacturing, and the production of solar-based clean fuels (such as hydrogen and other energy carriers) [10].

There are typically two types of CSP plants: line-focusing systems and point-focusing systems. Line-focusing systems concentrate the sun along a focal line, have a single-axis tracking system, and include parabolic troughs and linear Fresnel systems. Point-focusing systems concentrate the sun on one focal point (provides for higher concentration), have two-axis tracking systems, and include solar dish systems and solar tower technology. Parabolic trough, LFC, and solar tower systems provide power in the range of 10–300 MWs, while dish Stirling systems are deployed at much smaller scale (10–25 kW). While parabolic trough plants are the most commercially deployed CSP technology, they are still not fully mature, and require further cost reductions. Nearly all parabolic trough systems currently deployed do not have thermal energy storage and only generate electricity during daylight hours [9]. Thus, they rely on combustible fuel as a backup to firm capacity. For example, all CSP plants in Spain utilize natural gas for 12–15 percent of their annual electricity generation [10].

The direct normal irradiance (DNI)[1] at a location is the primary regional driver of CSP deployment, as it determines the technical feasibility of a CSP system. The technology is also highly centralized (except Stirling dish systems), and thus requires infrastructure to move to the electricity demand centers [11]. Only strong direct sunlight can be concentrated to the temperatures required for electricity generation by CSP systems, thus, CSP technologies are limited to hot, dry regions. For economic feasibility, these systems currently need DNI levels of at least 2,000 kWh/m^2/year. Higher levels of DNI have a considerable impact (not necessarily a direct impact) on levelized cost of electricity (LCOE), meaning that in general, LCOE levels are lower when the technology is deployed in regions with higher DNI levels. North Africa, Middle East, Southern Africa, Australia, the Western United States, and parts of South America have very high DNI levels suitable for this technology [9]. The regions with abundant solar resources that are appropriate for CSP-based electricity generation also tend to be remote regions, separated by large distances from demand centers, thus providing a challenge in expanded development and deployment [10].

1. *Direct normal irradiance (DNI):* The amount of solar radiation received per unit area by a surface that is perpendicular (or normal) to the solar rays. The amount of irradiance annually received by a surface can be maximized by keeping it normal to incoming radiation.

Parabolic Troughs

These systems are mainly comprised of parabolic-shaped mirrors, heat receivers, and supporting structures. The mirrors serve to concentrate the sun's rays onto a receiver tube at the focal line. Typically systems are oriented north-south, and ~~a single-axis tracking system tracks the sun east to west,~~ in order to maximize the amount of energy collected. The receiver tube at the focal line holds a heat transfer fluid to absorb the solar energy and transfer it to the steam generator (or a heat storage system, if one is being used). Typically the heat transfer fluid used has been synthetic oil, capable of holding temperatures up to 400°C. Newer trough designs utilize molten salt, which is capable of handling temperatures up to 540°C, and can be used either for heat transfer and/or as the thermal storage medium, improving the thermal storage performance greatly. Parabolic troughs have been the most widely deployed CSP technology. In the United States, a series of parabolic trough-based CSP systems, called the Solar Energy Generating Systems (SEGS-I to SEGS-IX) was built in the Mohave Desert of southern California between 1985 and 1991 [9]. Figure 3.4 displays a general schematic of a typical parabolic trough system.

Linear Fresnel Collectors (LFCs)

LFCs are similar to a parabolic trough, but use a series of long flat (or slightly curved) mirrors placed at different angles arranged in a field. A fixed receiver is positioned a couple of meters above the field of mirrors. Each line of mirrors is equipped with a single-axis tracking system and is optimized, ensuring that sunlight always concentrates on the fixed receiver. The focal line of LFCs has a secondary reflector to refocus the rays missing the tube, or several parallel tubes that form a multitube receiver wide enough to capture most of the focused sunlight without needing a secondary reflector. LFC systems are able to use cheaper flat glass mirrors and require less steel and concrete to build. The wind loads on LFCs are smaller, which results in a more stable structure, as well as less mirror damage (optical losses and breakage). Moreover, the mirror surface per receiver

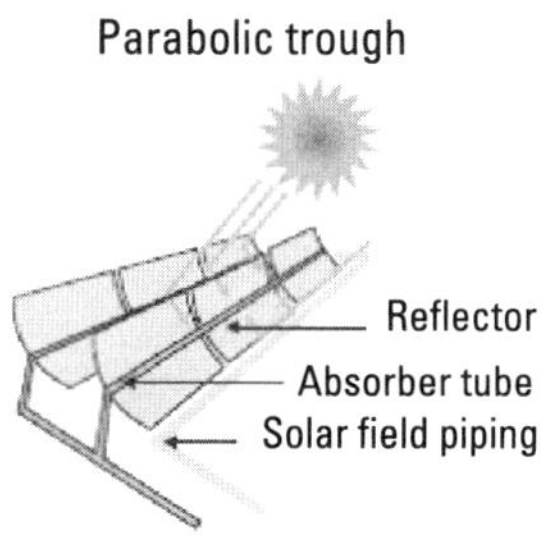

Figure 3.4 Parabolic trough system [10].

is higher in LFCs than in parabolic troughs. However, the optical efficiency[2] of LFC systems is lower than that of parabolic trough systems [9]. As LFCs use two parallel receivers for each row of mirrors, they need less land than parabolic troughs to produce a given output. The simple design of flexibly bent mirrors and fixed receivers of LFCs allows for direct steam generation, thus eliminating the need for (as well as the cost of) heat transfer fluids and heat exchangers. This in turn translates into lower investment costs [10]. Figure 3.5 displays a general schematic of a typical LFC system.

Solar Towers

Solar towers consist of a field of mirrors that focus the sun's rays onto a receiver mounted high on a central tower. The light reflected onto the receiver heats a fluid, the heat then produces steam to generate electric power. The mirrors used in the field are called heliostats, are computer-controlled, and track the sun individually in two axes. Solar towers are capable of achieving higher temperatures than parabolic trough LFC systems, ranging from 250°C to 1,000 °C (future plants), with about 600°C the average seen for current designs using molten salt. Solar towers are typically capable of generating electrical output on the scale of 10–50 MW. As the desired electrical output increases, the size of the field of mirrors required also increases. This, however, also leads to a greater distance between the receiver and the mirrors on the outer edge of the field, resulting in optical losses and slight errors in mirror tracking. The higher temperatures achievable by solar towers allow greater efficiency of the steam cycle, reduce water consumption for cooling the condenser, and allow greater temperature differential in the storage system (reducing costs or allowing greater storage for the same cost). Solar towers show good potential for increased application in the future, as they are capable of much higher temperatures (through

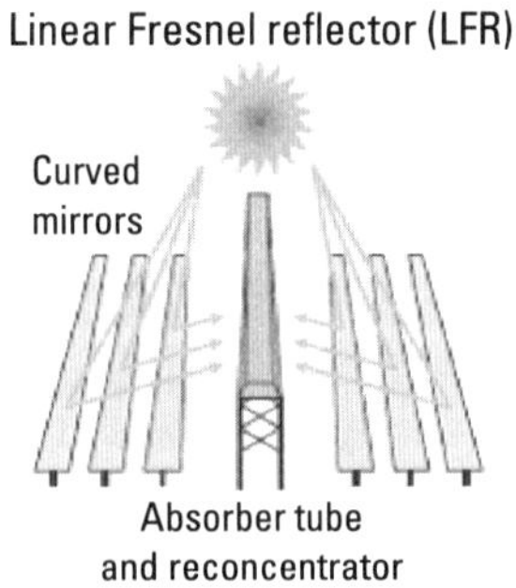

Figure 3.5 Linear Fresnel concentrator [10].

2. *Optical Efficiency* = Total thermal power absorbed by Receiver/(Direct Normal Irradiance × Total Mirror Aperture Area).

the use of molten salt as a heat transfer fluid/storage medium) and steam cycle efficiencies, thus reducing the cost of thermal energy storage. This will allow meeting peak electricity demands (especially for air conditioning) in an economic fashion, as well as load in the evening when the sun is not shining, in hot and arid areas with good solar resources [9]. Figure 3.6 displays a general schematic of a typical solar tower system.

Stirling Dish Engines

These systems consist of a parabolic, dish-shaped concentrator (similar to a satellite dish) that reflects the sun's rays onto a receiver (a Stirling engine or a microturbine) at the focal point of the dish. The concentration onto a single focal point can result in very high temperatures. Stirling dish systems track the sun via two axes, and have the highest efficiencies of all CSP systems. One advantage is that these systems are very modular and suitable for distributed generation. As Stirling dishes use dry cooling, they do not need large cooling systems, allowing for renewable electricity production in regions which are water-constrained. The most water-constrained regions also usually have the most plentiful solar resources. The smaller footprint and self-contained design of these systems allow them to be placed on slopes or uneven terrain, unlike other CSP technologies. Even though the levelized cost of Stirling dishes could likely be higher than other CSP technologies, their scalability and smaller size could allow them to meet an economically valuable niche in many regions, and could become an alternative to solar photovoltaics in arid regions. Other than higher costs, another drawback is that dish systems cannot easily use storage [9]. Figure 3.7 displays a general schematic of a typical Stirling dish engine system.

3.2.2 Impacts

The environmental and cost impacts and trends of solar technologies are described in the following sections.

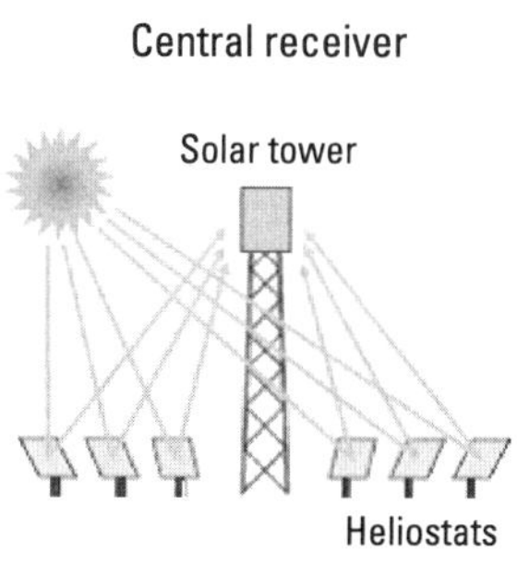

Figure 3.6 Solar tower [10].

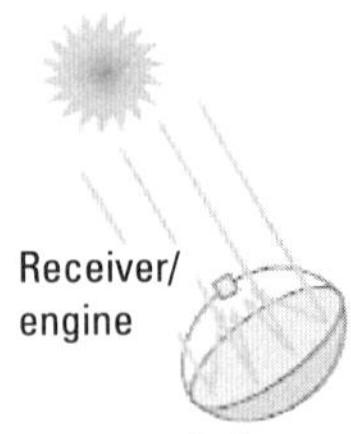

Figure 3.7 Stirling engine dish [10].

3.2.2.1 Environmental Impacts: Concentrating Solar Power

CSP technologies require water for cooling and condensing processes, and these requirements are relatively high, on the order of about 3,000 L/MWh for parabolic trough and LFC systems (which is a quantity similar to the water needs of a nuclear power plant). This compares to around 2,000 L/MWh for a coal plant and only 800 L/MWh for combined-cycle natural gas plants. However, solar tower systems require less water per MWh than trough plants, and Stirling dishes are cooled by the surrounding air and do not need any cooling water [10].

The water needs of CSP systems pose an important challenge to the use of CSP, as the optimal siting for these plants is generally in arid regions. Dry cooling (using air instead of water) is an alternative approach, but it increases the cost of electricity produced by 10 percent and reduces electricity production by 7 percent. This penalty is lower for solar towers than for parabolic troughs. An even more optimized approach is the use of hybrid wet/dry cooling systems. As water cooling is more effective but more costly, dry cooling may be used only in the winter when cooling needs are lower, and combined wet and dry cooling may be used during the summer. This approach may allow up to 50 percent reduction in water consumption with only a one percent drop in annual electrical energy production for parabolic trough systems [10].

3.2.2.2 Cost Impacts: PV

PV technologies have been demonstrating rapid declines in costs.At the beginning of 2012, thin-film module prices had fallen below $1/Watt (in the range of $0.84–0.93/Watt), and costs for c-Si modules were in the range of $1.02–1.24/Watt. PV module costs have a learning rate of 22 percent, which means that their costs will decline by about a fifth with every doubling of capacity. While utility-scale systems that use thin-film amorphous silicon, CdTe, and CIGS PV modules generally have lower capital costs compared to residential systems, this condition may not necessarily hold true if tracking is included in the costs [4].

Germany has the lowest PV system costs in the small-scale (<5 kW) residential market, with an average cost of $3.8/Watt for c-Si systems, while in Italy, Spain, Portugal, and the United States, costs for the same technology and market have been higher, in the $5.7–5.8/Watt range.

Even though the costs of PV systems are coming down at a rapid rate, the levelized cost of electricity (LCOE) of PV systems is still relatively high. For residential systems without storage, LCOE was $0.25–0.65/kWh in 2011. With storage, these costs increase to $0.36–0.71/kWh. In some cases the LCOE may be competitive with residential rates, especially if a region has adequate solar resources and high grid electricity rates for residential customers. Moreover, PV systems coupled with storage is almost always cheaper than diesel generators in off-grid electricity. For utility-scale thin film PV systems, the LCOE was $0.26–0.59/kWh in 2011 [4].

3.2.2.3 Cost Impacts: Concentrating Solar Power

While the operation and maintenance (O&M) costs of CSP plants can range between $0.02/kWh to $0.035/kWh, further cost reduction opportunities are available. The LCOE from CSP plants is also high at the moment, with $0.20/kWh to $0.36/kWh LCOE for parabolic trough plants, $0.17/kWh to $0.29/kWh for solar towers. These values may go as low as $0.14/kWh to $0.18/kWh in areas with abundant solar resources. It is expected that with capital cost reductions in the range of 10 percent to 15 percent, along with some reductions in O&M costs, by 2015, the LCOE of parabolic trough plants could decline to between $0.18/kWh and $0.32/kWh, and the LCOE of solar tower plants could decline to between $0.15/kWh to $0.24/kWh. By 2020, declines in capital costs of 28 percent to 40 percent are also likely to be reached [9].

The key parameters that determine the LCOE of CSP plants include the initial investment cost (including site development, components and system costs, assembly, grid connection and financing costs); capacity factor and efficiency; local DNI at the plant site; and O&M (including insurance) costs. The LCOE of CSP plants is closely tied to the DNI. The estimated LCOE of a CSP plant is expected to decline by 4.5 percent for every 100 kWh/m^2/year that the DNI exceeds the minimum required levels of approximately 2,000 kWh/m^2/year [9].

Parabolic trough CSP plants that do not include thermal energy storage have capital costs in the range of around $4,600/kW. However, these plants also have lower capacity factors (20–25 percent). If six hours of thermal energy storage is added, these lower capacity factors may be doubled, but capital costs will also increase, to between $7,100/kW to $9,800/kW. Solar tower plants have costs in the range of $6,300/kW to $10,500/kW, including energy storage of between 6 and 15 hours, and their capacity factors can reach as high as 40–80 percent [9].

3.2.3 Market Status

For all types of solar technologies, the United States is only a small fraction of the world market, while Germany is the strongest market for PV and Spain is the top market for CSP [6].

3.2.3.1 United States

PV capacity installed in the United States during the second quarter of 2012 was 742 MW, which represented a 45 percent increase over the previous quarter of 2012 and a 116 increase from the year before. The second quarter of 2012 was the largest quarter ever for utility PV installations. Over 20 deployments were realized, with a total capacity of 447 MW. Cumulative installed PV capacity reached 5,161 MW, and cumulative concentrating solar capacity (CSP and concentrating PV [CPV]) reached 546 MW by the end of the second quarter of 2012 [12].

Solar-based installations in the United States follow an overall trend of being moderate and consistent, with the exception of individual quarterly booms driven by utility installations. During the second quarter of 2012, while the residential market grew incrementally, the nonresidential market displayed large contractions, and the utility market expanded significantly. In fact, it is estimated that if the utility market had remained flat quarter-over-quarter, total installations would have actually declined by 18 percent instead of expanding by 45 percent. Over the upcoming few years, while all market segments are expected to display growth, utility installations are projected to continue to experience the greatest growth [12].

A total of 354 MW of CSP capacity were installed between 1985 and 1991 in California, and these facilities have been operating commercially since then. Between 1990 and 2000, there was a gap in interest and deployments of CSP, but interest has picked-up once again over the past ten years as a result of declines in investment costs and LCOE, along with some supportive policies [9].

3.2.3.2 Worldwide

It is projected that by 2050, PV technologies have the potential to provide approximately 11 percent of global electricity production and to avoid 2.3 Gt of CO_2 emissions per year [13].

Globally, Spain and the United States lead the CSP market, with the two countries making-up 90 percent of the global installed CSP capacity. Tax incentives and renewable portfolio standards in the United States and feed-in tariffs in Spain have led to these positions. Other emerging markets include Algeria, Egypt and Morocco, where CSP plants that are integrated with natural gas combined-cycle plants are being built. Australia, China, India, Iran, Israel, Italy, Jordan, Mexico, and South Africa have plans to develop large-scale CSP systems in the near future, and the United Arab Emirates has already started construction on a project. While at the end of 2010, about 1.2 GW of CSP capacity was online globally, by the end of March 2012, the global installed capacity had increased to around 1.9 GW. In terms of specific CSP technologies, at present, nearly all (94 percent) of installed CSP plants are based on parabolic trough systems, with an overall capacity of around 1.8 GW [9].

By 2013, the global capacity of CSP is expected to double for the second time in five years. However, the rapidly declining PV prices are posing a threat to growth in the CSP market. The track record of PV systems is also more established and familiar to investors, causing them to be favored. However, the bigger scale of CSP facilities (providing lower costs due to economies of scale), potential of using the technology in hybrid systems with fossil-based plants, and utility-scale energy storage capabilities of the technology may be factors of CSP to overcome PV's market dominance [11].

3.2.4 Challenges and Potential Solutions

Solar technologies are confronted with various technical, economic, and social challenges. For PV technologies, the main technical barriers include low conversion efficiencies of PV modules, performance limitations of system components such as batteries and inverters, and inadequate supply of raw materials such as silicon. The two main technical barriers related to solar thermal applications involve the heat carrying capacity of the heat transfer fluids and thermal losses from storage systems. The economic barriers related to solar technologies mainly relate to initial system costs. Moreover, these technologies are still considered high-risk by financial institutions, as solar projects have a shorter history, lengthy payback periods and a small revenue stream. Solar technologies also face institutional barriers, relating to factors such as limited capacities for workforce training, coordination of financial incentives, and insufficient numbers of trained workers to install and maintain solar energy systems [5].

For CSP systems, the main areas where cost reductions could be achieved include the following components [9]:

- *Solar field:* Mass production, cheaper components, and improvements in design may help reduce costs.
- *Heat transfer fluid:* Heat transfer fluids capable of higher temperatures will help to improve storage possibilities and reduce costs.
- *Storage system:* This is closely tied to the heat transfer fluid, as higher temperatures, especially from solar towers, can reduce the cost of thermal energy storage.
- *Balance of costs:* Including project development costs.

3.2.5 Emerging Opportunities

Third-generation PV systems are still under demonstration and have not yet been widely commercialized. This category includes CPV, dye-sensitized solar cells (DSSC), organic solar cells, and other novel solar concepts. CPV has the potential to have the highest efficiency of any PV module, but it is not clear at what cost premium.

Dye-sensitized solar make use of photo-sensitized components. These systems use low-cost materials and are simple to manufacture, but their performance can degrade over time with exposure to UV light and the use of a liquid electrolyte may cause freezing-related problems in cold weather conditions. Commercial efficiencies achieved are in the range of 4–5 percent, which is low, mainly because there are very few dyes that can absorb a broad spectral range.

Organic solar cells consist of organic or polymer materials and are inexpensive but not very efficient (4–5 percent range). A major challenge for organic solar cells is their instability over time. Organic cells may be applied to plastic sheets in a method similar to the printing and coating industries, allowing them to be lightweight and flexible and making them suitable for mobile applications and for fitting to a variety of uneven surfaces (useful for building integrated PV applications). The technology also uses nontoxic materials and is based on a very scalable production process.

3.3 Wind Energy

The energy of the wind is a vast resource that may be harvested at various scales and locations.

3.3.1 Basics

The basic concept behind the operation of a wind turbine is the transformation of the kinetic energy of wind into mechanical power, through the use of blades that provide via a drive shaft the mechanical energy to power the generator in the wind turbine.

3.3.1.1 Resource Base and Characteristics

The two main parameters that need to be considered in siting a wind turbine (or wind farm) are wind speed and the quality of the wind. To aid in the evaluation of the wind resource, wind power is categorized into seven classes, based on wind speed as m/s and wind power density as W/m^2. Table 3.1 displays the seven wind power classes. The roughness of the landscape where the turbines are to be located is another important consideration. Roughness may also be explained in classes (from zero to four), which depict the connection between the

Table 3.1
Classification of Wind Resources [3]

	At Height of 10m		At Height of 50m	
Wind Power Class	**Wind Power Density (W/m²)**	**Wind Speed (m/s)**	**Wind Power Density (W/m²)**	**Wind Speed (m/s)**
1	0	0	0	0
1–2	100	4.4	200	5.6
2–3	150	5.1	300	6.4
3–4	200	5.6	400	7.0
4–5	250	6.0	500	7.5
5–6	300	6.4	600	8.0
6–7	400	7.0	800	8.8
7	1,000	9.4	2,000	11.9

speed of the wind and landscape conditions. A class zero roughness indicates water surfaces and open areas with smooth surfaces, while class 4 indicates large cities with tall buildings [3].

3.3.1.2 Land-Based Wind

The harvesting of land-based wind may be through large, utility-scale turbines connected to the electric grid, or through much smaller turbines that may or may not be connected to the electric grid.

Utility-Scale Wind

The development of current wind technology may be traced back to 1979, with the mass production of wind turbines by various Danish manufacturers. These turbines started at 10–30 kW scale, and have led to the development of the multimegawatt modern wind turbines in use today. Currently, grid-connected wind turbines have capacities that average at around 1.16 MW, but many recent projects have been using turbines with capacities of 2–3 MW, or even larger (5 MW) [14].

Wind turbines are generally designed as either horizontal axis or vertical axis turbines. The design used most commonly in today's wind farms is based on three blades rotating around a horizontal axis. Even though vertical axis turbines are in the market, they do not hold a significant share of the utility-scale market, as they are less thermodynamically efficient [14].

In wind harvesting, the amount of available energy is a function of the cube of the wind speed, thus, every increase in wind speed has multiplying effects on the amount of energy that may be harvested. Capacity factors of wind turbines increase rapidly as the average mean wind speed increases. A doubling of wind speed translates into the increase of the power output of the wind turbine by a factor of eight. Moreover, at higher wind speeds and greater turbine heights, the wind blows in a more consistent fashion. The temperature of the air, as well as air turbulence, are also contributing factors. Colder (denser) air provides more energy, and turbulent air reduces the power output, while also potentially straining the equipment [14].

The swept area of the rotor of a wind turbine determines the maximum energy than can be captured by a wind turbine. By just doubling the rotor diameter, the swept area, and therefore power output may be increased by a factor of four [14].

The annual capacity factor of a wind facility is a key indicator of performance. As they gain experience with these facilities several countries have been reporting real-word data on wind farm capacity factors, in the 25 to 40 percent range. The annual capacity factor is defined as the amount of energy the facility generates over the year, divided by the amount of energy that would have been produced if the facility had been running at full capacity during that same time interval. Capacity factor for wind turbines is typically reliant on factors such as the quality of the wind resource, the availability (reliability) of the turbine itself to generate when there is enough wind, and the accuracy of the nameplate capacity rating versus rotor size. It should also be noted that capacity factor will be reduced if electricity production is curtailed by the utility, which is often the case during excess wind generation periods [15].

The main components of a typical wind turbine are [14]:

- *Blades:* Turbine blades are typically manufactured from fiberglass-reinforced polyester or epoxy resin, but newer materials such as carbon fiber are beginning to be used in order to allow for the high strength-to-weight ratio required by the larger turbines.
- *Nacelle:* This is the main structure of the turbine, where the main turbine components are contained.
- *Rotor hub:* The hub is usually attached to a low-speed shaft connected to the turbine gearbox.
- *Gearbox:* This component is housed in the nacelle, but newer direct drive designs do not require one. It converts the low-speed, high-torque rotation of the rotor to high-speed rotation with low-torque for input to the generator.
- *Generator:* The generator is also housed in the nacelle and converts the mechanical energy from the rotor to electrical energy.
- *Controller:* This component is the electronic system that monitors and controls the turbine, while collecting operational data. In addition, the presence of a yaw mechanism ensures that the turbine is always facing the wind.

- *Tower:* This is the main structure that holds the wind-harvesting gear at specified heights.
- *Transformer:* This unit is generally inside the tower of the turbine, and allows for the increase of the medium-voltage output from the generator, based on the requirements of the electric grid.

Small Wind

Small wind turbines provide clean power for homes, farms, small communities, and businesses, and are suitable for use on areas of land as small as one acre. Most small wind systems that are 100 kW or less typically have rotor diameters less than 50 feet, and are typically installed on towers less than 160 feet tall [16]. An average home in the United States would need approximately 5 kW of small wind capacity to meet all of its electric needs, although this value would range from 2 kW to 10 kW, depending on factors such as energy use and wind speed. While the payback time of installing a small wind turbine depends on various factors such as the turbine, the quality of wind at the installation site, prevailing electricity rates, and available financing and incentives, it usually varies from 6 to 30 years. The purchase and installation of a system that is capable of supplying electricity to an entire home has an average cost of $30,000, but the price can range from $10,000 to $70,000 depending on system size, height, and installation expenses. However, federal and state incentives in place can help bring down these costs [17].

A range of sizes and styles having been developed for small wind turbine systems (similar to large turbines), but horizontal axis small wind turbines dominate the current market at 95–98 percent of the market. Small wind turbines are sited close to demand centers, where wind speeds are often low and turbulent due to trees, buildings and other infrastructure in the vicinity. Thus, designing small wind turbines that will operate effectively in these environments has been challenging This has led to the push to use vertical axis turbine designs. Vertical axis turbines are not as sensitive to turbulent air conditions, have lower installation costs, require lower wind speeds to generate electricity (increasing capacity to serve areas with lower than average wind speeds), and have lower noise and vibration levels. However, these benefits come with the trade-off of lower efficiency [14].

There are three main growth drivers for small wind turbines: adequate wind resources, financial incentives (especially a suitable feed-in-tariff), and high electricity prices. While large wind farms have the flexibility of being located at sites with good wind resources (though not necessarily near the demand centers as the electricity generated is transmitted via the grid), small wind systems need to find locations that have a good wind resource and are the demand centers. Thus, these systems are generally located in rural areas and farms.

In today's market, distributed renewable energy technologies like small wind turbines rely heavily on financial incentives and support mechanisms to be feasible. Currently, in most markets, PV technologies are receiving more financial support than small wind turbines. In addition to favorable wind resources and financial incentives, grid parity is also essential for the success of small wind turbines. The higher the conventional grid electric rates, the better for small renewable energy systems, as their payback times are reduced and return on investment becomes greater. If a technology needs price supports over a long time period, it essentially becomes unsustainable in the long-term. Currently, small wind systems are actually closer to grid parity than small distributed PV systems, and are even closer to grid parity compared with electricity generated by diesel or in remote markets [18].

3.3.1.3 Offshore Wind

The offshore wind industry started with the installment of the first wind farm in 1991, off the coast of Denmark [19]. The force of wind in the lower parts of the atmosphere (where humans and other natural beings reside), is affected by the roughness of the surrounding area, and slows down due to the resulting friction. However, in offshore environments, friction is low, which enables stronger winds, and thus creates a bigger potential for energy to be harvested from wind. While land-based wind resources have been utilized for quite some time, the realization of the greater offshore potential with progress in related technologies led to the initial offshore wind projects to be deployed in the 1990s, with considerable growth in the following years [20].

It has been estimated by the U.S. National Renewable Energy Laboratory that the potential for offshore wind power in the United States is four times greater than the country's current total generating capacity from all sources [21]. The 28 states of the United States that are on ocean or lake coasts with potential for offshore wind development also represent the high loads that offshore wind can feed, with these states consuming 78 percent of the country's electricity [22]. Strong and constant winds offshore correspond well with these times of high electricity demand. Offshore winds blow strong especially during warmer days, when electricity demand is high due to air conditioning needs.

In offshore wind deployment, the depth of the water has an impact on the type of technology utilized. In shallow water (0–30 meters), monopoles and gravity foundations are used. In transitional waters (30–60 meters), tripods, jackets and truss-type towers can be used. In deep water (greater than 60 meters), floating structures, instead of fixed bottom foundations, may be required, but these structures are still in early stages of development [23].

In the United States, the East coast and the Gulf of Mexico have extensive areas of shallow water that extend far from shore. The West coast and the coast off Hawaii moves rapidly into the deep water category. The Great Lakes region, Lake Erie and portions of Lake Ontario have shallow waters.

The distance of an offshore wind deployment from the shore is also an important parameter, as it has various impacts, including the deployment's visibility from the shore, determination of whether the deployment is located in state or federal jurisdiction, and potential costs due to the length of underwater cable needed to connect the electricity tramsimission and distribution (T&D) systems on land [23].

3.3.2 Impacts

The social issue of perceived noise from wind turbines, the cost, and trends in wind technologies are described in the following sections.

3.3.2.1 Noise Impacts

Since wind farms are installed in various locations, with some close to populated areas, there has been opposition by some parties, including complaints about the noise from these farms. Current wind turbine designs have been effective in reducing any mechanical sounds that may be emitted from the wind turbines. However, the aerodynamic sound, often described as a "whooshing" sound, is what is normally recognized and disturbing to some. As wind farms are sited where the wind speed is higher than average, the background sound of the wind itself will often be more than the sounds that might be produced by operating these turbines. In fact, the noise level of wind turbines are in the range of the level of background sound in a typical home. Daily sources of noise we are subject to, such as the office environment, industry, and airplanes, have much higher noise levels [24].

3.3.2.2 Cost Impacts

The specific cost components of a typical wind turbine include: the rotor blades, the tower, and the gearbox, accounting together for about 50–60 percent of the turbine cost. The turbine itself accounts for 64–84 percent of the total installed cost. The cost contribution of these components will vary according to where the turbines are installed. The towers of a wind turbine, in general made from steel are a relatively mature component, not leaving much room for cost improvements. However, with increased competition, integration of lightweight materials, and more distributed location of manufacturers as markets expand-tower costs could decrease 15–20 percent by 2030. Using more carbon fiber in turbine blades and improving blade designs may help reduce their cost 10–20 percent by 2020. Gearboxes may also see costs decrease 15 percent by 2020, especially with the increasing share of gearless drive generators [14].

In 2010, wind farm installations in China and Denmark demonstrated installed costs for land-based wind farms in the range of $1,300/kW to $1,400/

kW. However, other major markets displayed costs between $1,800/kW and $2,200/kW. Installed costs for offshore wind systems are typically more expensive, costing $4,000/kW to $4,500/kW. The wind turbines themselves typically account for 64–84 percent and 44–55 percent of the total installed costs for land-based and offshore wind systems, respectively. O&M costs can make-up 11–30 percent of a land-based wind farm's LCOE, averaging $0.01/kWh to $0.025/kWh in major wind markets. For offshore wind systems, O&M costs are higher, and range between $0.027/kW and $0.048/kWh. The LCOE from wind farms varies depending on the wind resource and project costs, but wind represents a very competitive option, compared to other renewables. Assuming a cost of capital of 10 percent the LCOE of land-based wind farms in 2010 () ranged from $0.06/kWh to $0.14/kWh, and offshore systems were between $0.13/kWh and $0.19/kWh [14].

3.3.3 Market Status

The global drivers for wind power development tend to vary according to the priorities of specific regions and countries. For example, in India and China the drivers are strong economic growth and the need for electricity; in Europe, the driver is strong political support for managing global warming; and in the United States, global warming and security of supply are the drivers [25].

3.3.3.1 United States

The American Wind Energy Association (AWEA) conducts regular analyses of the wind industry in the United States. Their analysis of the third quarter of 2012 has underscored the growing U.S. wind industry [26]. Approximately 1,833 MW of wind power was put online during the third quarter of 2012, bringing the total cumulative U.S. wind power capacity to 51,630 MW. Approximately 40 percent more capacity was installed during the first three quarters of 2012 than the first three quarters of 2011. Installation activity was seen in 15 states with Kansas accounting for the most new installations, followed by Oregon, Texas, Oklahoma, and Nevada. The turbines installed during the third quarter of 2012 had an average capacity of 1.92 MW. While Texas has led by a wide margin in terms of capacity installed through 2011, Kansas picked-up the pace with the most capacity added during the third quarter of 2012.

In the United States, more than 19 MW of small wind turbines were installed in 2011, while revenues were $115 million. The cumulative installed U.S. small wind capacity increased to 198 MW in 2011 [16].

3.3.3.2 Worldwide

Evaluations conducted by the World Wind Energy Association (WWEA) [27] have shown that global wind installed capacity has reached approximately 250

GW (as of June 2012). This demonstrated a 16.4 percent increase compared to the previous year. The primary wind markets are in China, the United States, Germany, Spain, and India, and collectively they represent 74 percent of the global installed wind capacity. China represents the largest market, adding 5.4 GW of capacity in just six months, and reaching an overall installed capacity of about 67.7 GW in 2012. Most of the European wind markets displayed stronger growth in the first half of 2012 than in the same period of the previous year. The wind market in Europe is led by Germany, with a new capacity of 941 MW and a total capacity of 30,016 MW. Some new markets are also emerging in Eastern Europe, with 33 percent growth in Romania, 32 percent growth in Poland, 64 percent growth in Ukraine, and 64 percent increase in Latvia. The wind industry in the United States added about 2.9 GW between January and June 2012, which was 28 percent more than in the same period in 2011. However, future growth prospects are somewhat uncertain, as the industry relies on the Production Tax Credit (PTC), which is set to expire at the end of 2012. The two biggest Latin American markets, Brazil and Mexico, had modest growth rates in 2012, but were still above the global average. Australia also demonstrated 17 percent growth compared to 2011. The global wind power industry has grown at an average 27 percent per year rate between 2000 and 2011 [14].

Globally, more than 330 manufacturers offer wind turbines up to 100 kW, and the top five countries by number of small turbine manufacturers are the United States, China, Germany, Canada and the United Kingdom. It is predicted that the global small wind market will increase from 95 MW in 2011 to 700 MW in 2020 (corresponding to about 25 percent per year), with growth likely in the grid-connected distributed generation markets and driven by incentives (tax credits, rebates and feed-in tariffs). Japan, in fact, is expected to offer a significant feed-in tariff for small wind in 2012 as focus in the country shifts away from nuclear power and towards renewable energy. Off-grid small wind markets are also expected to grow, but at a slower pace. Utilization of small wind in the telecommunications industry is a growing market, where wind and solar hybrid power systems have been shown to lower operating costs by up to 90 percent, and it is estimated that this market could be worth up to $8 billion, with major deployments in Africa and Asia [16].

3.3.4 Challenges and Potential Solutions

The PTC, an incentive provided to generators based on the amount of renewable energy they generate in a year, has played an important role in the U.S. wind industry, with capacity increases following the expiration or extension of the credit. To quantify these impacts in detail, analyses were conducted [28] based on two scenarios: expiry of the PTC at the end of 2012, and a four-year extension of the PTC through 2016. For each scenario, the capacity of wind installed, jobs created, economic impact, and environmental impact were evaluated. It was seen that if the PTC is set to expire at the end of 2012, the U.S. wind market is expected to contract significantly in 2013. Moreover, by 2013, annual wind installations will be 2 GW (down from over 8 GW in 2012); total wind-related jobs will drop by nearly half; and total wind investment will drop by nearly two-thirds. However, if the PTC is extended for another 4 years, the U.S. wind market is expected to grow through 2016, with annual installations in the range of 8–10 GW through 2016. In addition, total wind supported jobs and total wind investment will also grow. The extension has the potential to result in an incremental 20 GW installed, 169,000 job-years, and $28 billion incremental investment through 2016.

In terms of the small wind industry in the United States, several challenges still remain [16]:

- Competition from rapidly decreasing PV prices, as well as lower natural gas prices keep pressure on small wind turbines to reduce costs, while not sacrificing reliability or durability.
- A large number of utilities remain resistant to small wind interconnection and economics.
- Local permitting and zoning requirements may hinder deployment.
- Better performance prediction tools and more certified site assessors and installers are needed to demonstrate the performance of small wind turbines, in order for these systems to garner financial support and customer confidence.

3.3.5 Emerging Opportunities

Small wind and offshore wind technologies, often are not the first technologies that come to mind when discussing wind harvesting, are gaining more interest and showing developmental progress.

3.3.5.1 Community Wind Concept

There are several trends emerging in the small wind turbine industry. Globally, small wind is starting to be recognized by individual home or property owners as a viable alternative source of electric power. The concept of community wind is also developing, where wind generation assets are owned by a group of local residents (generally farmers, local businesses, or sometimes the municipality) The community creates a Limited Liability Corporation (LLC) that enters into a power purchase agreement (PPA) with the local utility to sell the power. This model is common in parts of Europe, and is now also emerging in rural, windy areas of the United States, especially in Minnesota and Iowa. Communities are also realizing the benefits of small distributed wind generation in supporting local economies, especially in rural or poorly developed regions [18].

3.3.5.2 The Developing Potential of Offshore Wind

Offshore wind turbines not only make use of the higher average mean wind speeds in offshore environments, but also provide the option to build larger turbines with larger rotor diameters, leading to more energy being captured. However, larger turbines also have more complexities related to their construction, transport, and deployment, while costing more. While the designs and characteristics of the initial turbines used for offshore wind applications were mainly based on the designs of land-based turbines, recent offshore turbines are being designed specifically for marine environments. Typical designs for offshore turbines need to be resistant to the strong offshore winds, as well as the corrosive marine waters. However, the higher costs of the specific foundations, as well as the extra protection measures required, increase the overall capital costs [14].

3.4 Geothermal Energy

The amount of heat energy that the core of the Earth contains (geothermal energy) is several times more than enough to fulfill the needs of the entire global population. This unlimited resource is based on heat from the creation of the planet billions of years ago, as well as heat from the radioactive decay of naturally occurring unstable isotopes. Even though this vast resource has always been available to mankind, it wasn't until the latter part of the twentieth century that technologies were applied to utilize the resource and it started playing a more prominent role in our energy mix [29].

3.4.1 Basics

A unique characteristic of geothermal energy is that geothermal heat displays diversity in depths and temperatures, thus making it a more flexible resource that may be used for different objectives. For example, the lower-temperature heat available at relatively shallow depths (about 3–10 meters) may be utilized for heating and cooling homes and commercial buildings, through the use of highly efficient ground source heat pumps. This application enables reduction in electricity demand, and can result in even larger savings when combined with efficiency measures. In areas that have modest heat flows, the resource may be used for direct applications in industry such as food processing, greenhouses, aquaculture, materials drying, paper manufacturing, replacing the use of fossil fuels. Where high-temperature geothermal resources are available at greater depths, the resource may be utilized for electricity generation. While low-temperature, shallow resources are available pretty much everywhere, deeper high-temperature resources are more restricted to specific locations. In the United States, suitable resources for geothermal electricity generation are limited to about 30 percent of the geographical area of the country. However, more advanced technologies are currently being developed worldwide to expand the geographical reach to increase the amount of the resource that may be utilized [29].

Geothermal power plants are considered baseload power plants, meaning that they are able to provide a continuous source of power, as opposed to other intermittent renewable resources like solar and wind. This is an important feature that differentiates this renewable resource. However, this characteristic also precludes the resource from being used for load following or for providing peaking power capacities [29].

A key activity in deploying geothermal-based electricity is the resource exploration process. A variety of methods have been developed to identify regions where a resource base of adequate conditions and characteristics exists (depth, temperature, permeability, fluid amount, etc.) to allow for economic development. Initially, the geographic information available, such as geological maps and drill records, are evaluated to determine if suitable geothermal resources may exist in a given location. Once a location is identified as a potential target, more detailed geochemical and technical investigations (such as tacking water samples, drilling thermal gradient well, remote sensing surveys, etc.) are performed to further assess the appropriateness of the location [29].

3.4.1.1 Resource Base and Characteristics

The U.S. geothermal industry began in the 1950s with the development of a series of geothermal power plants called The Geysers. Even though it was developed more than 60 years ago, the site remains a mainstay of the U.S. geothermal industry. The Geysers is the world's largest geothermal site, currently producing at 725 MW [30]. Geothermal development sites to note in the United States include [31]:

- *California:* California is the current leader among U.S. states in installed geothermal capacity with 2,615.4 MW. Geothermal energy accounted for 42 percent of the in-state renewable electricity generation in 2010. With strong support for the Renewable Portfolio Standard (RPS), geothermal development continues. 31 new geothermal power plants are planned resulting in capacity additions of 1,065.6 MW to 1,110.6 MW.
- *Nevada:* The state currently ranks second, with 469 MW of installed geothermal capacity. Nevada also has the largest number (59) of proj-

ects under development, with an estimated planned added capacity of 631.5 MW to 641.5 MW. A strong state RPS also signals the potential for continued geothermal development in the state.

- *Hawaii:* On the Big Island of Hawaii, the Puna Geothermal Venture plant has an installed capacity of 43 MW and operates at 38 MW, supplying around 20 percent of the total electricity needs of the island. There are three other projects under development for the Big Island and also Maui. With increased support fromthe state (Hawaii passed legislation in 2009 that expanded its RPS goals to require 40 percent of electricity generation from renewable resources by 2030) there is potential for more geothermal development. The state currently has 42 MW installed capacity and 11 projects under development for an estimated 60 MW of planned added capacity.
- *Idaho:* The Raft River plant has a nameplate capacity of 15.8 MW but generates around 11.5 MW. Expansions to this plant are planned, and 9 other projects are in the pipeline, for a future potential of 33.2 MW.
- *New Mexico:* In 2008, a 0.24 MW pilot project was in operation, but was then brought offline. However, the same location is planned to host the Lightning Dock plant, which is expected to reach 15 MW.
- *Alaska:* In 2006, Alaska installed the first low-temperature power plant, 225 kW, at Chena Hot Springs. The plant has expanded its capacity to 0.73 MW. As the state has renewable energy goals to obtain 50 percent of its electricity from renewable resources by 2025, streamlined geothermal permitting, and increased funding for renewables good development is foreseen. 6 projects are under development for an estimated 25.4 MW of planned added capacity.
- *Oregon:* While that state currently has only 0.28 MW of installed geothermal capacity, there is great development potential with 16 projects under development for an estimated planned added capacity of 107.5 MW to 109.5 MW.

3.4.1.2 Dry Steam Plants

Dry steam plants (Figure 3.8) use hydrothermal fluids that are primarily steam. Naturally occurring reservoirs full of pressurized steam are tapped, and the pressurized steam is then released through a turbine. The turbine then drives a generator that produces electricity. Because the pressure is run entirely from energy generated from the earth, no fossil fuels are needed and the need to transport and store fuels is eliminated.

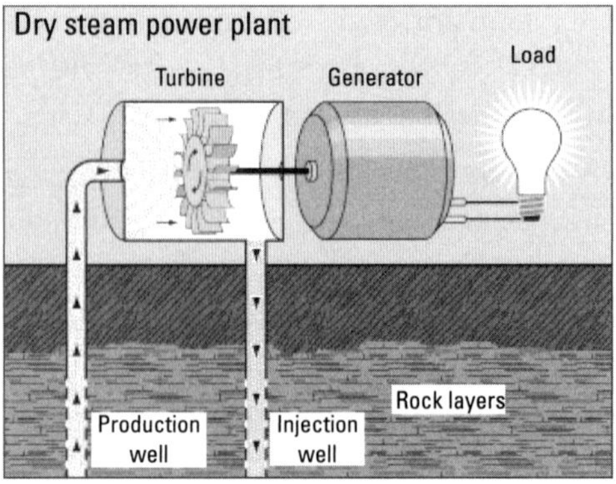

Figure 3.8 Diagram of a dry steam geothermal power plant [32].

This is the oldest type of geothermal power plant; it was first used at Lardarello in Italy in 1904 and is still very effective. Steam technology is used today at The Geysers in northern California, the world's largest single source of geothermal power [32]. The only emissions from dry steam plants are excess steam and marginal amounts of gases that escape from the underground reservoirs.

3.4.1.3 Flash Steam Plants

With few exceptions, flash steam plants (Figure 3.9) usually require temperatures in excess of 360°F (182°C) and are currently the most common type of geothermal power plant worldwide [32]. At these plants, fluid is sprayed into a tank held at a much lower pressure than the fluid causing some fluid to rapidly vaporize, or "flash." In a number of locations, this flashing occurs in the production well, whereas the vessels at the surface are used to separate the vapor and liquid phases of the fluid within the system. In either case, the pressurized

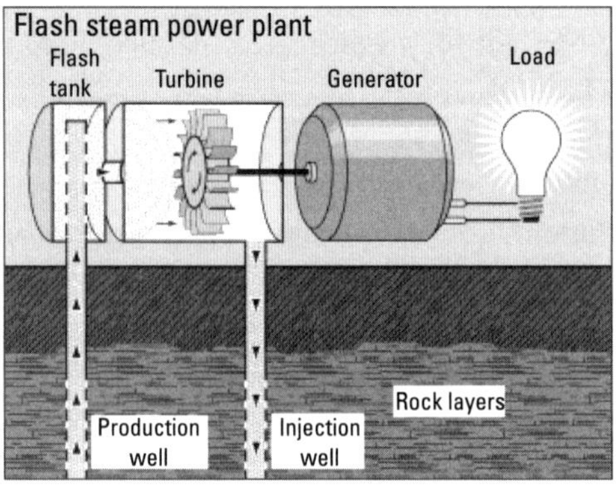

Figure 3.9 Diagram of a flash steam geothermal power plant [32].

steam vapor created by the flashing process is used to drive a turbine, which in turn drives a generator to create electricity. Any remaining liquid in the tank can be flashed again in a second tank to extract even more energy. Even after multiple flashes, less than 20 percent of the total liquid flow is flashed and utilized for energy production [32].

3.4.1.4 Binary Cycle Plants

If the temperatures are not consistently hot enough for the operation of a flash steam plant, a binary cycle power plant (Figure 3.10) can be used. Most geothermal areas contain water of moderate temperature (below 400°F) [32]. Energy is extracted from these fluids in binary cycle power plants. A hot, geothermal fluid and a secondary (i.e., "binary") fluid with a much lower normal boiling point than water pass through a heat exchanger. Heat from the geothermal fluid causes the secondary fluid to vaporize, which then drives the turbines. Because this is a closed-loop system, virtually nothing is emitted to the atmosphere. Moderate-temperature water is by far the more common geothermal resource, and most geothermal power plants in the future will be binary cycle plants.

Binary cycles will be used for lower temperature geothermal energy generation. Low-temperature geothermal energy is defined as heat obtained from the geothermal fluid in the ground at temperatures of 300°F (150°C) or less [32]. These resources are typically employed in direct-use applications such as district heating, greenhouses, fisheries, mineral recovery, and industrial process heating. However, some low-temperature resources can be harnessed to generate electricity using binary cycle electricity generating technology. A developing technology, the U.S. Geological Survey has identified more than 120,000 MW of untapped low-temperature geothermal resources in the United States [33].

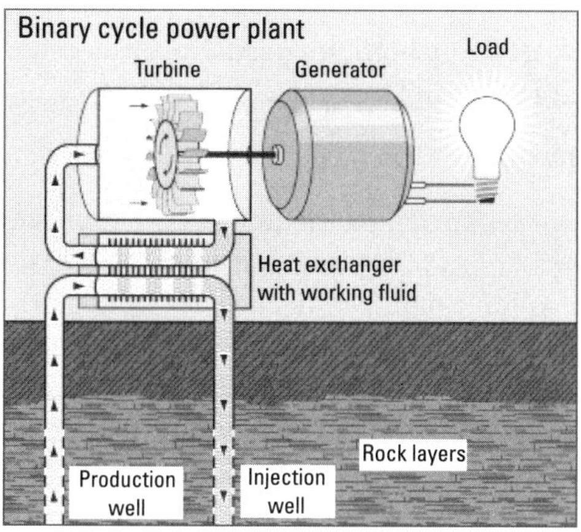

Figure 3.10 Diagram of a binary cycle geothermal power plant [32].

3.4.1.5 Geothermal Heat Pumps

While not considered a technology for use in the generation of power, the geothermal heat pump (also known as the ground source heat pump) is a highly efficient renewable energy technology that is gaining wide acceptance for both residential and commercial buildings. Geothermal heat pumps are used for space heating and cooling, as well as water heating. Its great advantage is that it works by concentrating naturally existing heat rather than by producing heat through combustion of fossil fuels.

The technology relies on the fact that the ground remains at a relatively constant temperature throughout the year, warmer than the air above it during the winter and cooler in the summer, very much like a cave. The geothermal heat pump takes advantage of this by transferring heat stored in the earth or in ground water into a building during the winter, and transferring it out of the building and back into the ground during the summer. The ground, in other words, acts as a heat source in winter and a heat sink in summer.

3.4.2 Impacts

The exploration, construction, and operation of a geothermal facility impacts the land. The construction phase creates the largest footprint, but once construction and testing are complete and operations have begun, these impacts are reduced significantly. Drilling a geothermal well may also require a drilling rig big enough to reach the needed depth to be brought to the site, which in turn might necessitate roads to be built, further impacting the land. However, even with these land impacts, geothermal technologies still have one of the lowest footprints of the various energy generation technologies [29].

3.4.3 Market Status

The use of geothermal energy both for electricity generation and in heat pump applications is gaining prominence.

3.4.3.1 Geothermal Power

The share of geothermal energy in the global renewable electricity mix is expected to increase over the next decade. This growth is initially expected in parts of the world where volcanic and tectonic activity are high (indicating potential for feasible geothermal resources), followed by areas where advanced drilling techniques may be used effectively to reach deeper resources. It is estimated that globally, a minimum of 190 GW of conventional geothermal resources which could be developed with current technology are available. It has been estimated that by 2050, approximately 160 GW of geothermal capacity will be installed globally [34].

Approximately 88 percent of the global installed geothermal capacity is present in only seven countries: the United States, Philippines, Indonesia, Mexico, Italy, New Zealand, and Iceland. Other countries with geothermal deployments include Costa Rica, El Salvador, Nicaragua, Japan, Kenya, Russia, Turkey, and Papua New Guinea. The global geothermal power market has experienced slowed growth over the past several years due to high upfront costs and long project development timelines, as well as the overall global economic downturn. However, the industry is getting back on track with renewed global financial support [34].

In terms of deployment of geothermal power, the United States currently is in the lead, with most resources concentrated in the Western part of the country and deployments led by California. The United States is followed by Philippines and Indonesia in terms of installed capacity [34].

Globally, while conventional (hydrothermal) geothermal projects are the basis of nearly all of the current installed capacity, small enhanced geothermal and coproduced systems are starting to enter the market. At the moment, enhanced geothermal projects are still in the demonstration phase, but if results of value are obtained from these projects, these technologies hold the potential to attract significant investment and make a difference in the geothermal market. Geothermal exploration technologies are also developing; supported by better methods, and streamlined land leasing approaches. Flash steam and dry steam technologies are the dominant technologies on the market, making-up 87 percent of online capacity. However, smaller binary technology with lower prices and suitability for lower temperatures are gaining traction. Even though geothermal technologies have high initial costs, their low emissions, stable electric capacities, and low levelized cost of energy allow the technology a significant place in the power market [34].

3.4.3.2 Geothermal Heat Pumps

The strong growth of GHP applications globally have been the main driver behind the increase in direct use geothermal applications. This has been led by the United States, China, and Sweden. In 2010, global geothermal direct use reached 51 GWth (a 79 percent increase over the direct use capacity of just five years before), with 78 countries using the method. GHPs make up 50 percent of the market, while bathing/swimming applications made-up 25 percent. The remaining 25 percent of market share included applications such as space heating, greenhouses, industrial process heating, aquaculture, snow melting/cooling agricultural drying, and other uses [35].

While GHPs make up most of the direct use capacity globally, they also are the fastest growing direct use application, displaying 18 percent CAGR This is expected to increase to a rate of 21.7 percent CAGR between 2011 and 2017. GHP technologies are technically deployable throughout the world, even if operational efficiencies differ (due to variations in soil conditions and temperatures). This greatly increases their potential for growth. The United States, China, Sweden, Norway, and Germany currently collectively account for 77 percent of global GHP capacity [35].

3.4.4 Challenges and Potential Solutions

The siting process for geothermal energy development, involving leasing and permitting, may be lengthy. Exploration and drilling activities remain expensive and risky, as properly identifying potentially productive geothermal reservoirs is difficult. The relatively small size of the industry, in combination with the exploration risks, makes it challenging to obtain appropriate financing. The greatest quality and potential geothermal resources are generally found in more remote areas not too close to demand centers. Thus, more transmission and distribution lines are needed, adding cost.

3.4.5 Emerging Opportunities

Coproduced hot water, enhanced geothermal systems, and mineral recovery from geofluids are three emerging pathways that enhance the geothermal system's value.

3.4.5.1 Coproduced Hot Water

Much of the development within the geothermal industry may also be tied to oil and gas production. For example, coproduced hot water is a by-product of oil and gas wells, and 25 billion barrels are produced each year in the United States. Historically, this hot water has been an inconvenience and a disposal issue; however, it is now being looked at as a resource to produce electricity for field use or for purchase by the grid. Coproduced geothermal resources have the potential to help diminish GHG emissions and extend the economic life of oil and gas fields [36]. Coproduced, low-temperature power production can bring a source of renewable energy to an area that lacks many renewable resources—the eastern Gulf Coast. The area from Louisiana eastward to Florida is rich in carbon-intensive resources while relatively absent of wind and water resources. Furthermore, solar resources are not on a scale large enough to be used for commercial production. Coproduced power production allows what was previously considered to be waste to be used for profit, while also mitigating the production of carbon-intensive oil and gas.

3.4.5.2 Enhanced Geothermal Systems

Exploring some geothermal reservoirs may not be economical due to lack of water and/or permeability. This is where engineered reservoirs, or enhanced geothermal systems (EGS) may be created to produce energy in a more economical fashion. EGS technology has the potential to allow access to the substantial levels of heat resources located at depth. It is estimated by the DOE that the application of EGS technology is capable of providing at least 100,000 MW of electricity within 50 years [37]. The DOE has funded seven EGS projects under FY 2008 and ARRA appropriations. These demonstrations were selected to become the first EGS power plants in the United States, ranging in location from Alaska, down through Oregon and California, and into Nevada and Utah. These projects are intended to advance EGS technologies with the potential to produce geothermal power nearly anywhere in the United States [31].

3.4.5.3 Mineral Recovery from Geofluids

Mineral recovery from geofluids is a purely value-added technology, applicable wherever subsurface fluids are being processed. Ore minerals in these geofluids are already in the aqueous phase, which reduces processing time, energy, and costs, making this technology's commercial viability potentially significant. From silica recovery alone, there is a potential of $10.2 million/year revenue from the Salton Sea 50 MWe power plant [38]. Two of the more valuable metals contained in geothermal fluids, lithium and manganese, are expected to increase in demand as reusable batteries and hybrid vehicles become more mainstream. A significant barrier to large-scale mineral recovery operations has been limited advancement of extraction technologies from the laboratory scale to pilot or field scales.

3.5 Biomass Energy

The generation of electricity from biomass resources provides several benefits. Biopower represents one of the only base load renewable options and is able to utilize feedstocks that have widespread availability. Biomass represents the fourth largest global energy resource after coal, oil, and natural gas, estimated at about 14 percent of global primary energy. By 2021, forecasts indicate that the use of biomass for commercial power generation applications could reach 115 GW, representing $138 billion in cumulative investment [39].

A key characteristic related to biomass is its availability upon demand, as energy is stored within the biomass until it is needed. Other forms of renewable energy, such as solar or wind, are dependent on variable environmental conditions such as sunlight intensity or wind speed. Biomass technologies are highly scalable, thus they can be used to create power for a wide range of applications from farms and remote villages to sizable cities [40]. Cofiring biomass resources in conventional fossil-based plants is a good option for the near-term, as much of the existing power plant equipment can be used without major modifications, and can be less expensive than building a new biopower plant. In addition, cofiring with biomass allows a portion of the fossil resources used to be replaced, and results in reductions in sulphur dioxide, nitrogen oxides, and other harmful air emissions. Once the initial integration is made, there is little or no loss in efficiency from adding biomass [40].

The future potential of biopower will likely depend on the ability of governments around the world to implement carbon regulations, carbon trading, renewable portfolio standards, and other incentives. The European Union has incentivized biopower investments through its Emissions Trading Scheme, but the failure other regions to provide similar incentives has limited more robust biopower expansion [39].

3.5.1 Basics

The main processes used to convert biomass resources into electricity may be categorized as thermal-chemical processes (combustion, gasification and pyrolysis) and bio-chemical processes (anaerobic digestion).

3.5.1.1 Resource Base and Characteristics

Unlike wind, solar, and hydro, biomass-based electricity generation requires a feedstock that needs to be produced, collected, transported and stored. The economic feasibility of biopower is thus dependent on the availability of a secure, long-term supply of an appropriate biomass feedstock at a competitive cost. Biomass feedstocks may be based on urban or rural sources. Urban sources include urban wood waste (such as packing crates and pallets), biogas from wastewater treatment, landfill gas, municipal solid waste, and residues from food processing. Rural sources include forest residues, agricultural residues (such as corn stovers and wheat stalks), energy crops, and biogas from livestock effluent. The ash and moisture content, and the homogeneity of biomass resources have an important impact on the cost of biomass feedstock per unit of energy, transportation costs, pretreatment and storage costs, and also determines the appropriateness of different conversion technologies [41].

Table 3.2 provides the typical feedstock cost and plant capacity ranges for the different types of biomass feedstocks utilized [42].

In terms of the bulk densities and energy densities of various biomass resources compared to each other and compared to coal none of the biomass resources are inherently as energy dense as coal. Thus biomass requires a range of pretreatment and upgrading processes to enhance handling, transport, and conversion activities [42].

3.5.1.2 Direct Combustion

In the combustion process, a conventional Rankine cycle is used. Biomass is burned (or oxidized) in a high-pressure boiler to generate steam, with efficiencies of around 23 percent to 25 percent. The exhaust of the steam turbine may be condensed to produce power or used for another heating activity [41].

Table 3.2

Biomass Feedstocks, Costs, and Biopower Plant Capacities [42].

Biomass Feedstock				
	Wastes	**Processing Residues**	**Locally Collected Feedstocks**	**Internationally Traded Feedstocks**
	Organic waste (MSW) Sewage sludge Manure/dung	Timber residues Black liquor Bagasse Rice husks Food-processing wastes	Agricultural residues Forestry residues Roundwood and thinnings Energy crops	Roundwood Wood chips Biomass pellets Biomethane Pyrolysis oil
Typical Feedstock Costs ($/GJ)	Negative to zero	0–4	4–8	8–12
Typical Plant Capacity (MWe)	0.5–50	0.5–50	10–50	>50

Combustion of biomass for power generation is a mature commercial technology, and may be applied at scales of a few MWs to hundreds of MWs. Globally, over 90 percent of the biopower generation utilizes this method. The main components of a combustion-based biomass facility are the biomass-fired boiler that produces steam, and the steam turbine used to generate electricity. The two main types of boilers used are stoker and fluidized bed boilers. Stoker boilers burn fuel on a grate and produce hot flue gases that are then used to produce steam. The ash from the combusted fuel is removed in a continuous manner by the fixed or moving grate. Fluidized bed boilers suspend fuels on upward blowing jets of air during the combustion process [41].

Biomass may also be cofired with coal in a coal-fired power plant, where a percentage of biomass is added to the fuel mix in the coal-fired plant. Cofired could be up to 5–10 percent of biomass (in energy terms) and 50–80 percent with extensive pretreatment of the feedstock with only minor changes in the handling equipment. If cofiring involves using biomass at percentages above 10 percent, or if biomass and coal are burning separately in different boilers (parallel cofiring), then changes in the process equipment are required [41].

Cofiring of biomass with coal is becoming increasingly common practice, with approximately 55 GW of coal-fired capacity currently cofired with biomass in North America and Europe with varying degrees of biomass integration (3–95 percent). The incremental investment costs of biomass cofiring are relatively low, but the process can increase the cost of a coal-fired power plant by as much as a third. However, cofired facilities allow higher efficiencies to be achieved [41].

3.5.1.3 Pyrolysis

Pyrolysis can be considered a subset of the gasification process. The pyrolysis process is similar to gasification, but here the partial combustion is stopped at a lower temperature (450°C to 600°C), resulting in the creation of a liquid bio-oil that is used to generate electricity, as well as gaseous and solid products. The bio-oil created is similar to crude oil, but is contaminated with acids and therefore needs to be treated [41].

3.5.1.4 Gasification

Gasification is the process where biomass is partially combusted in an environment containing low levels of oxygen, which results in the generation of a gaseous product called syngas, which is a mixture of carbon monoxide, water, carbon dioxide, char, tar and hydrogen, and can be utilized by various technologies to generate electricity (combustion engines, microturbines, fuel cells or gas turbines.) Fuel cells and turbines offer higher electrical efficiencies. The gasification is a two-step process where biomass is decomposed by heat, yielding mainly volatile materials. These volatile hydrocarbons are then gasified at higher temperatures in the presence of a reactive agent (air, oxygen, steam, or a mixture of these gases) to produce syngas, along with tar and ash. The process is endothermic, meaning that it requires heat to proceed, and the syngas generated will require a clean-up process to remove unwanted contaminants. Gasification technologies are commercially available, but more R&D is required to enable widespread commercial use. Key remaining challenges include improving fuel flexibility; removing particulates, alkali-metals and chlorine; and removal of tars and ammonia [41].

3.5.1.5 Anaerobic Digestion

The anaerobic digestion process involves decomposing any variety of biological material in a warm, wet, and airless environment to generate a gas (referred to as biogas) which consists mainly of methane and carbon dioxide. After clean-up, the biogas can be utilized in internal combustion engines, micro-turbines, gas turbines, fuel cells and Stirling engines for electricity generation. The method provides a very effective means to treat organic materials, including energy crops, residues and wastes from various industrial and agricultural processes, and municipal waste streams. Multiple feedstocks may be codigested to achieve the best balance of biogas yield and process stability [41].

3.5.2 Impacts

The life-cycle carbon impacts and other environmental impacts, as well as cost impacts of systems converting biomass resources to energy are detailed in the following sections.

3.5.2.1 Life-Cycle Carbon Impacts

An energy generation activity which is considered carbon-neutral is one where the carbon dioxide released and absorbed are in balance. In biopower production, there is no commonly accepted definition for a carbon-neutral activity. In fact, multiple statements have been made, including the following [43]:

- Biomass energy is carbon neutral because biomass is naturally carbon neutral.
- Biomass energy is neutral if the activity removes as much carbon dioxide as was emitted into the atmosphere.
- Biomass energy is neutral only if the net life-cycle emissions are zero.
- Biomass energy is neutral if it achieves lower net increases in atmospheric GHGs when compared to alternative energy activities.

Each of these statements is based on different assumptions, and there are pros and cons to each. For example, the claim that that biomass energy is carbon neutral due to the inherent qualities of the biomass feedstock does not account for GHG emissions released due to management of crops grown for energy production (such as the use of fertilizers.) The amount of carbon emitted from biopower generation is due to feedstock production (cultivation and harvest), feedstock transport, and the biopower technology type. However, the feedstock production step also absorbs carbon during growth [43].

The carbon neutrality for biopower is based on the carbon flux (GHG emission or sequestration) of several parameters [43]:

- *Feedstock type/ how the feedstock is managed and harvested:* The carbon flux depends on the type or mixture of feedstocks used. For example, GHGs may be higher from agricultural biomass due to the use of fertilizers. Or the removal of large amounts of woody biomass may reduce some carbon emissions that would have been released if the woody biomass was left to decompose in the forest.
- *Feedstock transportation method:* How far the feedstock needs to be transported determines the amounts of GHGs emitted.
- *Energy generation technology:* The type of technology, as the potential use of carbon sequestration methods will determine the carbon flux.
- *Timeframe to replenish the feedstock:* The timeframe assigned for biomass feedstock replenishment will determine carbon dioxide sequestration rates to balance out the GHGs emitted during biomass combustion.

Various life cycle analyses (LCAs) conducted on biopower pathways have demonstrated that while there is a significant variation in the GHG and energy balances, most pathways display considerable savings of GHG emissions compared to conventional fossil-based alternatives. These savings are between 63 percent to 99 percent compared to the use of coal, and between 37 percent and 98 percent compared to the use of natural gas [44]. While a full LCA is a detailed method to account for all steps and their impacts on the overall carbon balance, this evaluation method can be expensive and time-consuming to complete. Thus, bioenergy's carbon neutrality remains a contentious aspect of the bioenergy debate [43].

The lowest life-cycle GHG emissions can typically be achieved through the use of residues and wastes onsite (e.g., in pulp and paper mills), as use of wastes and residues avoids methane emissions that occur through decay of organic waste. Thus, emission savings of more than 100 percent, compared to fossil fuels, may be achieved. For cofiring and other utility-scale bioenergy generation from wood chips and pellets, GHG reduction potential is somewhat lower, but still sizeable compared to fossil fuels. Biopower production may be combined with carbon capture and storage, to ultimately achieve "negative emissions." Figure 3.11 compares the life-cycle GHG emissions (excluding those due to land use changes) from various biomass pathways and conventional fossil pathways, and demonstrates that significant savings may be achieved [42].

However, carbon dioxide emissions due to land use change caused by biopower production pathways may reduce or negate the GHG benefits gained. The extent of this effect depends on when and where the changes take place, as well as how the respective carbon stocks in the form of standing biomass and soil carbon and emission cycles are modified when managed for bioenergy

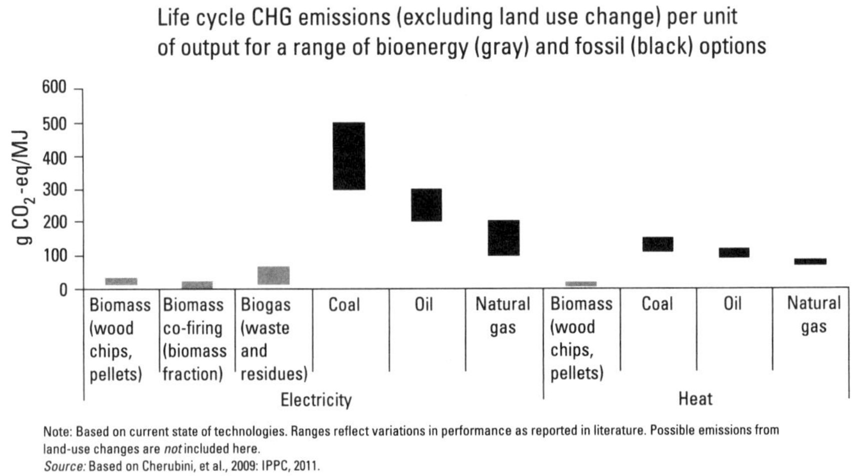

Figure 3.11 Life-cycle GHG emissions comparison for bioenergy and fossil pathways [42].

feedstocks compared to business-as-usual. It may take several decades to remove the atmospheric carbon dioxide that is released in the early phases of a biopower process. Thus, effective land use management efforts need to be put in place to reduce this initial large release of GHGs.

3.5.2.2 Other Environmental Impacts

Biomass-based electricity production has other environmental impacts, such as impacts on biodiversity, soil fertility, water quality, and the use of water. The overuse of natural resources through increased extraction rates of forest biomass result in a negative impact on soil quality, carbon stocks, and biodiversity. For biomass resources based on agriculture, issues of concern include excessive use of fertilizer and pesticides, as well as overuse of irrigation water. If land use is well managed and indigenous energy crops are used, most of these issues may be addressed. In addition, health impacts due to fine particles from biomass combustion may be an issue. Filters and scrubbers to remove these particles from conventional utility-scale power plants are rarely applied to smaller-scale biomass combustion units, but new technologies for pollutant removal at smaller scales are becoming available [42].

The steady growth in global population, along with economic growth in emerging economies is expected to result in 70 percent increase in global food demand and a net demand of an additional 70 Mha of arable land in 2050. Moreover, biomass demand in other nonenergy sectors needs to be met, while growing interest in other industries (e.g., chemicals) will likely further increase the overall demand for biomass in the future [42].

3.5.2.3 Cost Impacts

The installed cost of biomass power generation technologies varies according to the technology used and the country of deployment. If combined heat and power (CHP) is deployed, these types of plants have installed costs that are typically higher than electricity-only configurations. In 2010, the total installed costs of different biopower technologies were as follows [41]:

- Stoker boilers: $1,880–4,260/kW.
- Circulating fluidized bed boilers: $2,170–4,500/kW.
- Anaerobic digester power systems: $2,570–6,100/kW.
- Gasification technologies (including fixed bed and fluidized bed solutions: $2,140–5,700/kW.
- Cofiring biomass at low-levels in existing thermal plants: Added investment of $400–600/kW.
- Using landfill gas for power generation: $1,920–2,440/kW.

For biopower facilities, O&M costs have a significant impact, contributing to 9 to 20 percent of the levelized cost of energy (LCOE). This value may be lower for facilities utilizing cofiring, and greater for those that involve extensive fuel preparation, handling, and conversion [41].

The LCOE of biopower power plants ranges from $0.06/kWh to $0.29/kWh, depending on capital and feedstock costs. If low-cost feedstocks are available and capital costs are modest, biomass can be a very competitive power generation option. Even if feedstock costs are high, the LCOE range for biopower is still more competitive than for diesel-based generation, allowing biomass to be a good solution for off-grid or mini-grid electricity applications [41].

The feasibility of biopower plants is closely connected to the long-term availability and low cost of the feedstocks used. If wastes are usedwhich otherwise would have disposal costs or that are produced onsite at an industrial installation feedstock costs may be avoided. The transport of feedstocks is an important factor in biopower applications. If significant transport distances are involved, overall costs are impacted, as biomass resources typically have low energy densities. Capital costs for preparation and handling of biomass feedstocks can contribute to 6 to 20 percent of the total investment costs of the power plant for systems larger than 550 tons/day. If biomass-based energy efforts are integrated with traditional agriculture processes, such crop rotation and agricultural intensification, yields may be increased and price reductions realized. Relevant analyses have suggested that cost reductions of 2 to 25 percent may be achievable [41].

Predicting biomass cost reduction potentials is complicated by factors such as the local supply chain, resource potential, land availability, competitive industrial uses (e.g., biochemical), risks of deforestation, sustainability criteria, etc. Many biopower generation technologies (e.g., direct combustion in stoker boilers, low-percentage cofiring, anaerobic digestion, municipal solid waste incineration, landfill gas and combined heat and power) have already reached maturity and cost reductions through scale-up will be modest. Less mature technologies (atmospheric biomass gasification and pyrolysis) as well as ones that are only at the demonstration or R&D phases (integrated gasification combined cycle, biorefineries, biohydrogen), have the potential to realize significant cost reductions as commercial experience is gained. Gasification using wood or waste wood as feedstock may achieve capital cost reductions of 22 percent by 2020. Stoker direct combustion technologies may achieve reductions between 12 and 16 percent. Anaerobic digestion technologies could realize cost reductions on the order of 17 to 19 percent by 2020 [41].

3.5.3 Market Status

Current installed biopower capacity is distributed relatively evenly across North America, Europe, Latin America, and Asia Pacific. However, the Asia Pacific region is projected to attract one third of cumulative investment and to maintain the largest installed base of biopower capacity through 2021, due mainly to expanding use in China, South Korea, India, and the Association of Southeast Asian Nations (ASEAN) countries. The United States has great potential to integrate biopower resources, but insufficient policy support and cheap fossil fuels are projected to limit growth over the next decade. Developments in the production of advanced biofuels are expected to lead to competition for biomass feedstocks within the United States. At the same time, if ethanol production is expanded, it is projected that this expansion will have ripple effects in Brazil, likely driving expansion of biopower generation capacity, as sugar and ethanol processing facilities there coproduce heat and power from bagasse [39].

3.5.3.1 Feedstock Market

The biomass market is mainly focused on the trade of wood chips and pellets, but the majority of biomass feedstock is not traded, as it is used for other purposes such as domestic cooking, heating and lighting. Thus, most of biomass resources used for power generation come from nontraded sources, such as wastes and residues from agricultural and industrial processes, and forestry that are consumed locally. In certain regions significant commercial markets for biomass feedstocks may exist, in general, the feedstock is local in nature, meaning that biomass power generation plants are small in scale (typically up to 50 MW). Securing adequate amounts of low-cost feedstock for large-scale plants, taking transportation into account, is challenging. A small but growing trade is emerging; however, with regards to pellets and wood chips, and more supportive policies for renewable power generation may support further growth in these markets. Wood chips and pellets were initially used for district heating, but are now increasingly being used to cofire fossil fuel power plants or to displace them entirely. Canada, the United States, and Western Russia are the major exporters of pellets to Europe, and the largest consumers are Sweden, Denmark, the Netherlands, Belgium, Germany, and Italy. The Netherlands is the largest EU importer of wood pellets. Trade related to wood chips is larger than that of pellets and tends to be more regional and international, with Japan being the main market. China is emerging as an important forest biomass importer from North America, but the primarily use of this biomass is for heating and charcoal production since cofiring and direct burning of biomass is still in the early stages [41].

3.5.3.2 Biopower Market

It is projected that by 2050, biomass-based generation could provide 3,100 TWh of electricity globally, which would represent 7.5 percent global electricity generation, and 1.3 Gt of CO_2-equivalent emissions savings per year. Electricity supply from biomass resources has displayed a steady rising trend since 2000, and generation is still concentrated in OECD countries. However, China and Brazil are also becoming increasingly important producers, especially utilizing agricultural residues. Within OECD, European countries have demonstrated the largest growth [42].

3.5.3.3 Biogas Market

After nearly a decade of strong growth, current installed global production capacity of biogas is more than 800 billion cubic feet per year. The biogas market is a relatively small portion of the overall bioenergy market, but has displayed fast growth and reached $17.3 billion in global revenue in 2011. It is expected to reach $33.1 billion by 2022. Germany currently leads the biogas market, but its domestic market is becoming increasingly saturated, which is expected to lead firms to venture abroad seeking opportunities, and result in biogas use accelerating across Europe, Asia Pacific, and key developing markets. In the United States, although biogas potential is significant, the combination of low natural gas prices, unstable policy, and a challenging regulatory environment is expected to restrict industry growth over the next decade. Growth of the biogas market is expected to be the greatest in those countries where a combination of increased renewable energy demand, distributed infrastructure, stable policy, and waste management initiatives will drive increased investment. Over the next decade, Europe is expected to continue holding a large share of the market, and the Asia Pacific region is expected to increase its share [45].

3.5.4 Challenges and Potential Solutions

Biomass resources have several advantages over fossil fuels as electricity generation feedstocks. They are widely distributed; relatively easy to collect and use; can produce less net carbon dioxide emissions per unit of useful energy delivered when sourced sustainably; and usually contain less sulphur than coal or oil. However, the combustion characteristics of biomass feedstocks are quite different than those of fossil fuels, which results in some technical and economic challenges [42].

Biomass is the organic material of recently living plants, crops, as well as waste material (such as municipal solid waste), and thus results in a diverse resource base, with different chemical compositions and characteristics. This diversity may become an issue in utilization, as some technologies are not as flexible in accepting a wide range of feedstocks for operation. The general chemical composition of biomass resources consists of high (but variable) moisture content and fibrous structure (comprised of lignin, carbohydrates or sugars and ash). The chemical composition of the biomass feedstock also impacts its energy density. The moisture content of biomass can vary from 10 to 60 percent, or even higher for some organic wastes. When high moisture content biomass resource is used for anaerobic digestion, the energy value of the feedstock is reduced, which in turn increases transportation costs and the fuel cost on an energy basis. If various drying methods are utilized, the energy density of the feedstock may be enhanced, which helps reduce transportation costs and improves combustion efficiency. However, the tradeoff is that these additional steps increase feedstock prices, while the energy balance decreases significantly due to the energy consumption used for these pretreatment procedures. It is also important to note, though, that while the cost per ton of feedstock may be increased, in the end, the price of the feedstock per unit of energy may be improved [41].

The utilization of biomass resources also poses risks to [46]:

- *Food Security:* To meet the food requirements of the increasing global population, global food production will need to increase by about 50 percent by the year 2030, which in turn will require agricultural land use to be increased by 13 percent during the period. The expanded cultivation of energy crops will put an added pressure on the competition for agricultural land, further increasing food costs. Thus, it is predicted that the dynamics of energy markets and food prices will be closely linked in the future.
- *Biological Diversity:* The increased cultivation of energy crops will place an increased demand on agricultural products, and this will likely be met by intensifying existing agricultural systems, or claiming new agricultural land, all at the expense of biological diversity.
- *Climate Change Mitigation:* As natural ecosystems are converted to agricultural land to meet bioenergy needs, greenhouse gases are released.

3.5.5 Emerging Opportunities

Integrated biorefineries and waste-to-energy systems are emerging as comprehensive and valuable solutions.

3.5.5.1 Integrated Biorefineries

Biorefineries are emerging as integrated facilities capable of producing a variety of fuels and other products from a certain feedstock. The operation becomes economically competitive through the production of high-value coproducts, in addition to comparably low-value bioenergy, including biofuels. Biorefineries process various types of biomass feedstocks into energy and a range of marketable products, such as food, feed, materials, and chemicals. These facilities contribute to the sustainable and efficient use of biomass resources by way of generating a range of products to different markets [42].

Some of the key challenges related to the development and commercialization of biorefineries relate to financing, market viability, and feedstock diversity. Integrated biorefineries need to optimize use of the biomass resource to create a range of products that are matched to market demand and can also compete with fossil fuels. As biorefineries are able to utilize a variety of biomass feedstocks, it is possible to capitalize on each specific region's geography and climate. However, this diversity also creates a challenge in trying to develop replicable feedstock supply systems and specialized conversion technologies. Figure 3.12 demonstrates the main components of an integrated biorefinery [47].

3.5.5.2 Waste-to-Energy

Waste-to-energy (WTE) offers three key benefits in managing waste: reduction of waste volumes by at least 90 percent, recovery of metals and other materials, and the generation of renewable baseload energy. It is a biopower pathway that holds great potential but is greatly underutilized globally. As urban areas develop and urban populations increase, the wastes generated in these areas from daily residential, commercial and industrial activities also increases. In 2011, nearly 2 billion tons of municipal solid waste (MSW) was generated in urban areas around the world, and it is expected that this value will rise to at least 2.9 billion tons by 2022, representing an estimated 240 GW of untapped potential. Most of the trash discarded currently ends-up in landfills or open pits, and is not utilized for energy generation. In 2011, only 11 percent of the global MSW stream was utilized as a source of fuel in WTE plants. However, with the increasing demands and constraints of urban areas, WTE is reestablishing itself as an attractive technology option to promote low carbon growth [48].

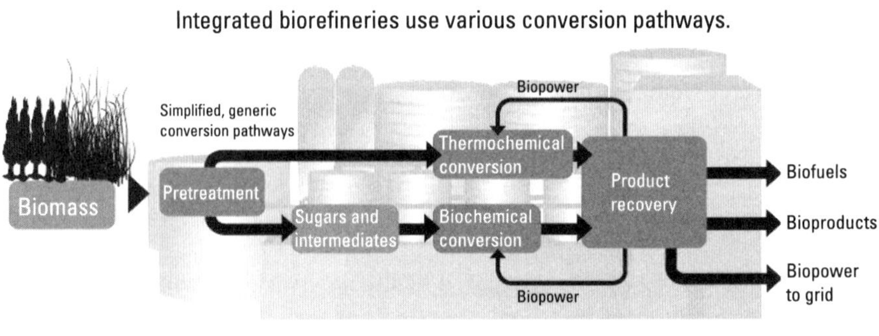

Figure 3.12 Schematic of an integrated biorefinery [47].

China recently surpassed the United States as the leading generator of MSW, and is either at or near capacity for many of its landfills. Based mainly on this trend in China, the Asia Pacific region has emerged as a strong player in the WTE field and is projected to capture half of the global market by 2018. The WTE market in EU market remains significant, but parts of the market are already saturated, and growth is confined to only a few of the member states. In the United States, after a 15-year pause, WTE activity is gaining momentum, but growth is expected to fall behind that of the Asia Pacific region and EU. If recent discoveries of abundant shale gas resources pan out, lower natural gas prices in the United States (and potentially Europe and China) may potentially inhibit more widespread WTE expansion. It is estimated that at a minimum, 261 million tons per year of WTE capacity will come online worldwide by 2022, and under an optimistic forecast scenario, this number could be as high as 396 million tons per year [48].

3.6 Water Power

The movement and heat of the vast water resources around the globe may be utilized via a wide variety of technologies to harvest clean energy. Some of these technologies are mature, while others are still on the research and development end of the spectrum.

3.6.1 Basics

Conventional large-scale and smaller-scale hydropower, as well as several marine hydrokinetic technologies are described in the following subsections.

3.6.1.1 Conventional Large-Scale Hydropower

Hydropower operates on the principle of transforming the potential energy of a mass of water flowing in a river or stream with a certain vertical drop (also called "head"). The extent of electricity generation of a hydropower plant is proportional to the head and flow of water. These facilities use the energy potential of the flowing water to turn a turbine, which then provides the mechanical energy to drive a generator to produce electricity.

Renewable electricity generation from hydropower facilities dates back many decades. For example, it contributed to 40 percent of the power generated in the United States in 1920. As many cities and industries were located near rivers, hydropower was able to supply electricity close to the load centers and was even termed "white coal" back then. Hydropower is not considered variable in the same sense as wind power or solar PV, as it has inherent storage capabilities and its generation is more predictable, even for run-of-river plants. However, hydropower may be considered variable over longer time scales, as it depends on precipitation and water run-off. Hydropower facilities also contribute to efforts related to climate change adaptation, as regulation of water flow is an important component of these efforts. Reservoirs can regulate water flows for freshwater supply, flood control, irrigation, navigation services and recreation [49].

There is no standard definition of hydropower sizes, but in general, the following descriptions are used [50]:

- *Large Hydropower:* 100 MW or more of capacity feeding into a large electricity grid.
- *Medium Hydropower:* From 20 MW to 100 MW almost always feeding a grid.
- *Small Hydropower:* From 1 MW to 20 MW usually feeding into a grid.
- *Mini Hydropower:* From 100 kW to 1 MW that may either be standalone, as part of a mini-grid, or connected to the grid.
- *Micro Hydropower:* From 5 kW to 100 kW, providing power for a small communities or remote areas away from the grid.
- *Pico Hydropower:* From a few hundred Watts up to 5 kW, often used in remote areas away from the grid.

There is no agreed definition for small and large hydropower, and the description differs from country to country [50].

Hydropower facilities may also be categorized as follows [50]:

- *Run-of-River:* These hydropower projects have no (or very little) storage capacity and generation is dependent on the timing and size of river flows.
- *Reservoir:* These systems are able to store water behind the dam in a reservoir, which may have small or very large storage capacities.
- *Pumped Storage:* These systems typically utilize off-peak electricity to pump water from a reservoir in order to generate at peak times and provide grid stability and flexibility services.

Hydropower plants typically have very long lifetimes in the range of 30 to 80 years. With regular upgrades to the electrical and mechanical systems but no major upgrades to the most expensive components (dams and tunnels), hydropower systems can last for years. The capacity factor achieved by hydropower projects has some considerations that vary from other renewable electricity generation technologies, as a hydropower has significant flexibility in its design. These systems may be designed to have a high installed capacity and low capacity factor, providing electricity mainly to meet peak demands and to provide ancillary services to the electric grid. The systems may also be designed to have a lower installed capacity and higher capacity factor, with potentially less flexibility in generation to meet peak demands and provide ancillary services. Data obtained from various project around the world have demonstrated this variability in design and capabilities, indicating that hydropower systems may have capacities ranging from as low as 23 to as high as 95 percent, with an average capacity factor of 50 percent [50].

The main components of a hydropower facility are [50]:

- *Dam:* The dam that holds back the water, creating a large water reservoir that can be used as storage. The dam may also include a desilter to manage sediment build-up behind the dam.
- *Intake, Penstock, and Surge Chamber:* When gates located on the dam open, gravity conducts the water through the penstock (a cavity or pipeline) to the turbine. A surge chamber (tank) is used to reduce surges in water pressure that could be damaging to the turbine.
- *Turbine:* The typical turbine type used today is the Francis Turbine, which allows a side-by-side configuration with the generator.
- *Generators:* As the turbine blades turn, the rotor inside the generator also turns to generate electricity.
- *Transformer:* The transformer increases the voltage for more efficient (lower loss) long-distance transport of the electricity through the transmission lines.
- *Transmission Lines:* These lines are used to transport the generated electricity to a grid-connection point or to the user directly.
- *Outflow:* The area where used water is carried out through pipelines and reenters the river downstream. A spillway may also be used here, where the water is allowed to bypass the generation system and be "spilled" in times of flood or very high inflows and reservoir levels.

While the operation of hydropower is carbon-free, GHG emissions do occur as a result of construction activities, silting in the reservoirs, and decomposition of organic matter (mainly an issue in tropical regions). Hydropower facilities also tend to have large spatial footprints, and impacts on the local populations, ecosystems, and biodiversity, which need careful evaluation when developing a hydropower project [50].

Some remaining areas for development relating to hydropower technology include development of variable-speed generation technology, efficient tunneling techniques, integrated river basin management, hydrokinetics, silt erosion resistant materials, and better management of environmental issues [50].

3.6.1.2 Micro Hydropower

Micro hydropower systems may be categorized as low-head and high-head systems. Low-head systems have a vertical drop of less than 5 feet, and also have short pipelines that allow the water to drop through the runner.[3] High-head systems have a vertical drop of over 10 feet, and consist of hundreds of feet of pipeline to head and water delivered to the runner via several nozzles [51].

The basic components of a small hydropower system are: diversion and intake screen (directs water from into the pipe or channel); penstock/pipeline (transports water to the turbine); turbine (generates electricity and includes nozzles, runner, and generator); batteries and related electronics (stores energy and regulates turbine); dump load (absorbs excess energy); and transmission and distribution (delivers energy to its end use) [51].

Micro hydropower systems are capable of generating large quantities of energy from a small water flow, with minimal impact, and on a continuous basis. They can be beneficial to meet a home's energy needs and also be useful for off-grid systems. Especially if the hydro resource is located relatively close to the point of use, micro hydropower may be a more feasible renewable electricity generation option than other renewables. If appropriate rules and incentives are in place and the micro hydropower system is connected to the grid, excess power generated may also be sold back to the grid for extra income, and/or offer back-up power generation in situations of grid unavailability [51].

Micro hydropower systems are relatively small systems and are especially beneficial to users who are independent of the electric grid. Most micro hydropower systems fall under the "run of river" category, and thus do not have a storage reservoir. As only a portion of the stream flow at a given time is used to generate power, environmental impacts are minimized. Micro hydropower systems allow individual users to utilize a low-cost, renewable source of electricity that is generated via available technologies and is effective over a long time period [52].

3.6.1.3 Marine and Hydrokinetic Power

Marine energy technologies harvest energy from either the natural movement or temperature of water bodies, and represent a very large energy potential, but are still in the earlier part of their development path and not widely commercialized [53]. The industry is beyond the initial conceptual stage and is progressing,

3. *Runner:* Part of the turbine that receives the water and turns its energy into rotation in a shaft.

but at a relatively slow pace. At this point, if the industry does not deliver on its promise, there may be a risk of losing investor interest and political support. Different views support the notions that either the marine and hydrokinetic industry is currently where the wind industry was in the mid-1970s, or that it could be closer to the wind industry's status in the early 2000s, when wind deployment boomed. The United Kingdom currently leads in support and deployment of marine and hydrokinetic technologies, and several other countries with strong marine resources (United States, Australia, South Korea, Spain, and Portugal) have also made commitments to support this evolving industry [53].

Tidal projects currently comprise over 90 percent of marine kinetic capacity totals. The total global installed MHK capacity is projected to reach 5.5 GW by 2017, still dominated by tidal technologies [53].

As marine energy resources have very high energy content profiles, they have less capital cost per unit of electricity generated, and the high capital costs associated with renewable energy technologies is largely avoided with marine kinetics. However, these resources have uncertain O&M costs, in the order of approximately 40 percent of total project costs, as compared to 10 percent for solar and 20 percent for wind [53].

Tidal Energy

Tidal energy is created as a result of the relative motions between the earth, sun, and moon, as well as the Earth's rotation. Changes in gravitational forces due to this relationship incite variations of the sea level (and thus water currents) in specific cycles. Energy from the difference in high and low tide, as well as the tidal stream, may be harvested utilizing different technologies [20].

The technology of utilizing the tides of the ocean or rivers for tidal power generation shows similarities to conventional hydroelectric power generation. The main components of a tidal system are a dam, tidal pond (created by the dam), powerhouse (containing a turbo-generator), and a channel gate to allow for the flow of tide in two directions. During the flood tide, when the gate of the dam is opened, the tidal waters fill the tidal basin, and when the dam is filled to capacity, the gates are closed. Once the tidal waters have receded, the tidal basin is released through the turbo-generator. Power may be generated either during the flood tide (when the water level increases) or during the ebb tide (when the water retracts). It has been shown by various studies that the most economical application of tidal power is when the mean tidal range (vertical difference between high and low tide) is 16 feet or more [3].

An important benefit of tidal power generation is that tidal currents are regular, predictable, and not influenced by weather and climate conditions. However, the technology is still considered relatively new and further research, development, and demonstrations are required. The technology used to harvest tidal energy is similar to conventional wind turbines. As water is approximately 800 times more dense than air, tidal turbines are stronger and heavier (and also more costly) than wind turbines, but at the same time, are capable of capturing more energy [3].

While tidal power systems are nonpolluting and make use of a nonexhaustible resource, they also may have potentially negative impacts on the marine environments that they are built in. Some fish migrate between saltwater to freshwater, and when the channel gates of the tidal system are closed, this impacts their path and potentially lives. Also, dams may lead to less water flow between the sea and the basin, decreasing the average salinity in the basin, which in turn affects the food chain of organisms living in the basin. However, a dam may also at times offer protection for coastal areas by serving as a storm surge barrier during very high tides [3].

In the United States, the largest number of locations with high kinetic power density suitable for tidal energy are in Alaska, followed by, Maine, Washington, Oregon, California, New Hampshire, Massachusetts, New York, New Jersey, North and South Carolina, Georgia, and Florida Average tidal stream power densities exceed 8,000 W/m2) in several of these locations [54].

Wave Energy

Uneven heating patterns of the globe lead to winds that generate ocean waves, and the motion of these waves holds kinetic energy that may be harvested for clean electricity production through the use of various advanced technologies. The duration and speed of the wind, the area and depth of water affected by the wind, as well as seabed conditions all determine how much kinetic energy the waves hold. The total power available from waves globally is estimated at 2–3 million MWs [3].

The total available wave energy resource in the United States is estimated at 2,640 TWh/yr, but only about 1,170 TWh/yr of the total resource is theoretically recoverable, given technology limitations [54].

The energy contained in waves is dependent on the height and period (the time between successive peaks) of the waves. Technologies that convert wave energy to electricity may be categorized by their location in the sea. Offshore wave energy converters are designed for deep waters that are distant to the shore, while near-shore and shoreline systems operate in shallower water [55]. Harvesting technologies are either fixed to the ocean floor or float on the surface of the ocean, and may be deployed either close to the shore or in deeper waters. The energy from the movement of the waves may be captured using a variety of methods, such as floating devices that bend with the waves, or devices that make use of pressure fluctuations in tubes resulting from the movement of the waves [53].

Ocean Thermal Energy Conversion

Ocean thermal energy conversion (OTEC) technology relies on the temperature difference between the warmer water surface and the cooler water depths. OTEC technologies may be categorized into three main systems: open cycle, closed cycle, and hybrid cycle systems [53].

3.6.2 Impacts

Investment costs for hydropower projects have a wide range, due to the variations in the site, design choices and the cost of local labor and materials. Significant civil works required for hydropower project development result in a large contribution of the cost of materials and labor to the overall costs, especially as compared to other renewable energy projects. Average investment costs for large hydropower plants (that include storage) range from $1,050/kW to $7,650/kW, and for small hydropower projects the range is between $1,300/kW and $8,000/kW. If existing hydropower facilities are used with some additional capacity added, costs may be as low as $500/kW [50].

Annual O&M costs for large hydropower plants range from 2 to 2.5 percent of the investment costs, while for small hydropower plants, the range is between 1 and 6 percent, or even higher in some situations [50].

The levelized cost of electricity (LCOE) of hydropower is typically low, but at the same time, very site-dependent, ranging from $0.01/kWh for additional capacity at an existing hydropower project, to $0.05/kWh for a more extensive upgrade project. The LCOE for large hydropower projects typically ranges from $0.02/kWh to $0.19/kWh, while small hydropower LCOE values range from $0.02/kWh to $0.10/kWh [50].

3.6.3 Market Status

Hydropower currently generates about 16 percent of the world's electricity and comprises the majority (over four-fifths) of global renewable electricity. At the moment, more than 25 countries rely on hydropower for 90 percent of their electricity needs, while 12 countries are exclusively reliant on hydro. Canada, China, and the United States have the largest hydropower generation capacity [50].

The key markets for hydropower development reside mainly in Asia (led by China) and Latin America (led by Brazil), accounting for about 60 percent of global activity. The main focus in OECD, North America, and Europe is the enhancement of existing facilities, which leads to increased capacity, as well as new pumped storage facilities. Completely new hydropower capacity development is also occurring, but in relatively modest quantities. South America has 11 GW planned (with an additional 16.3 GW at the feasibility stage) and

Western Asia has a total of 15.5 GW of capacity under construction (mainly in India). Europe leads the hydropower market in the small hydropower category, and the technology serves as the second highest contributor to the European renewable energy mix. China has plans to start construction on 140 GW of capacity over the next five years, and in collaboration with Iran, also plans to build the world's tallest dam with a capacity of 1.5 GW in Iran's Zagros Mountains. In North America and Europe, even though new facilities are under construction, efforts mainly focus on enhancing existing plants and adding pumped hydro storage capacity [50].

As only about one quarter of the global hydropower capacity has been tapped into, the potential for growth is good, but challenges like long lead times; project design, planning and approval processes; and the time required to secure financing (especially for large-scale facilities) indicate that capacity growth is more likely to be slow and steady than rapid [50]. Untapped hydropower potential is considerable in Africa, Asia, and Latin America, as detailed in Figure 3.13. It is projected that by 2050, global hydropower capacity will double (up to almost 2,000 GW) and provide for over 7,000 TWh of electricity generation. A majority of hydropower capacity development is expected in large projects in emerging economies and developing countries, where such projects may improve access to modern energy services, relieve poverty, and

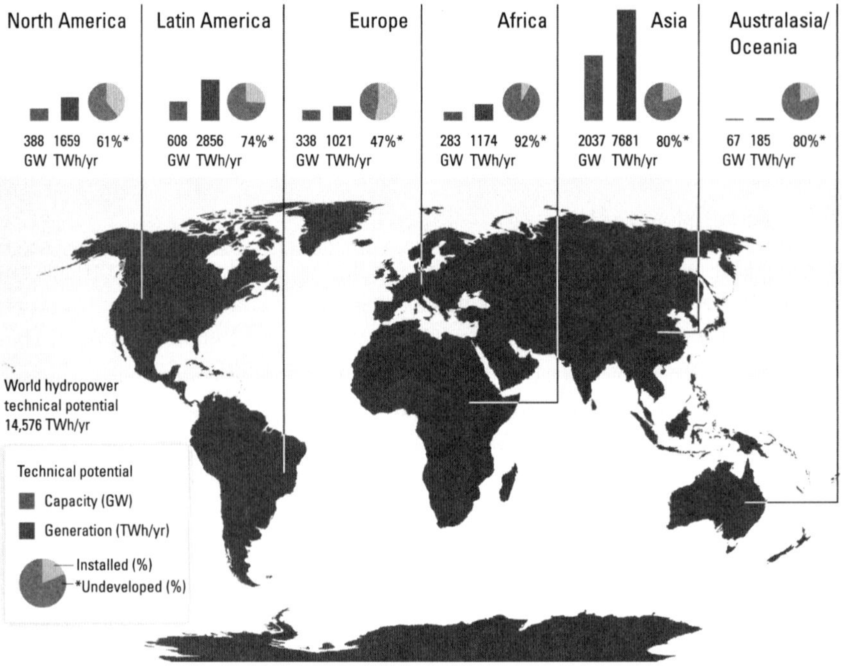

Figure 3.13 Untapped hydropower potential around the world [49].

assist the social and economic development of local communities. Hydropower development activities in developed countries are expected to involve upgrading existing plants, and may deliver benefits related to various energy-related issues that these countries face [49].

The average annual growth rate of hydropower is in the order of 2.5 percent per year, which seems small when compared to growth rates of wind and solar, but it has a much larger already existing base. It can be seen from Figure 3.14 that over the last decade, electricity generation from additional hydro capacities has kept pace with generation from all other renewables combined [49].

3.6.4 Challenges and Potential Solutions

Hydropower provides low-cost and competitive electricity, but significant initial investments in the range of tens of billions of dollars are required, making financing a challenge. Energy policies, especially in industrialized countries), tend to overlook hydropower, as it is generally believed that the economic potential for hydropower was exhausted decades ago, and/or that hydropower plants are damaging to the environment. Moreover, the potential for hydropower to act as energy storage tohelp balance fluctuations in electricity supply from wind and solar technologies is not clearly understood. However, while hydropower technologies are improving, the environmental, social and economic conditions of sustainability are also being taken into consideration. In today's hydropower technology landscape, the safety of dams is in general very high, and many opportunities exist to increase the capacity, efficiency and environ-

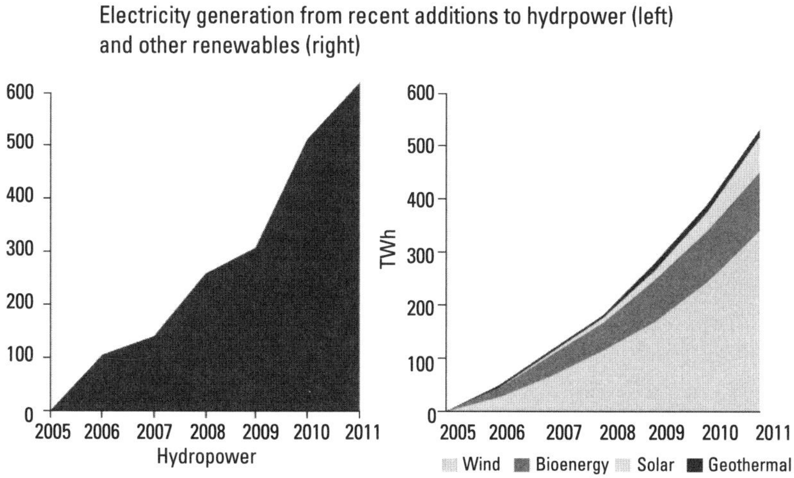

Source: IEA, 2012b.

Figure 3.14 Comparison of hydropower capacity additions vs. other renewables [49].

mental performance of older facilities, as well as many opportunities to build new ones [49].

The most important drivers for hydropower development in the future are expected to include the following factors [49]:

- Continuous local generation capability and low life-cycle costs;
- Reliability of electricity production;
- Safe operation;
- Environmental and socially sustainable development, as well as providing climate change mitigation;
- Energy services enhancing grid stability and enabling greater penetration of variable renewables;
- Large-scale energy storage for seasonal load balancing;
- Nonenergy services such as flood control, water supply, and irrigation. These services will be even more critical when freshwater needs and adaptation to climate change are considered.
- Upgrades, redevelopments and improvements to existing hydropower plants;
- Addition of new hydropower facilities to existing dams originally built to provide nonenergy services;
- Energy security through use of a local resource.

Wave and tidal technologies are in direct contact with the marine environment, which poses additional challenges. Devices used need to be more strictly designed to prevent water from entering them under the abrasive and high-pressure conditions of the marine environment. Bio-fouling. the accumulation of marine organisms on submerged structures, is another concern. Using some moving parts (with speeds at least 1 m/s), as well as using specific paints reduces these issues. Offshore devices must also be designed so as to resist the corrosive nature of the marine environment. Moreover, development of offshore electric distribution networks faces more challenges than onshore systems when it comes to permitting, as factors and regulations related fisheries, shipping, oil/gas exploration, and conservation areas also need to be evaluated [20].

One consideration for marine-based renewable technology development is the issue of water borders and rights. As the industry advances and the sites closer to the shores become fully utilized, developers will be in search of resources farther offshore, an issue already becoming a consideration in the UK, Norway, and Germany. At some point in the future, an offshore energy development will need to span national jurisdictions, which will raise more issues to consider and involve negotiations between nations. Each country has its own method of governing its surrounding waters. For example, Poland, Denmark, Ireland, and several other nations regard their territorial and exclusive economic zone (EEZ) waters as the same for permitting purposes. On the other hand, the United States gives control of water up to three nautical miles to the state, and those beyond that are managed by the Department of the Interior, as they are considered federal waters [20].

3.6.5 Emerging Opportunities

Conventional hydropower technologies are mature and thus do not have much room for improvement, but the hydro resource itself is still globally untapped to a great extent. Thus, opportunities exist for added renewable capacity.

Micro hydropower systems are serving the needs of expanding distributed renewable generation.

Technical barriers are amplified for wave and tidal technologies, as the technologies are not as mature. Thus, not many developers have adequate installation or operation experience. Currently, wave and tidal technologies under development are very diverse, and are expected to overcome some of these barriers and eventually converge into fewer specific designs, due to factors such as technology optimization (emergence of the most feasible designs) and commercial developments (bankruptcies and mergers).

References

[1] Heslin Rothenberg Farley & Mesiti P.C. (Cleantech Group), *Clean Energy Patent Growth Index (CEPGI): Second Quarter 2012 Results,* October 22, 2012.

[2] U.S. Department of Energy, *Office of Energy Efficiency and Renewable Energy, 2011 Renewable Energy Data Book,* 2011.

[3] Khaligh, A., and O. C. Onar, *Energy Harvesting: Solar, Wind, and Ocean Energy Conversion Systems,* Boca Raton, FL: CRC Press, 2010.

[4] International Renewable Energy Agency, Solar Photovoltaics, *Renewable Energy Technologies: Cost Analysis Series,* Volume 1: Power Sector, Issue 4/5, June 2012.

[5] Timilsina, G. R., L. Kurdgelashvili, and P. A. Narbel, *A Review of Solar Energy: Markets, Economics, and Policies, Policy Research Working Paper 5845,* The World Bank, October 2011.

[6] Sherwood, L., *U.S. Solar Market Trends 2011,* Interstate Renewable Energy Council, August 2012.

[7] Website: The George Washington University, *The GW Solar Institute, Frequently Asked Questions,* Available at: http://solar.gwu.edu/FAQ/solar_potential.html.

[8] U.S. Department of Energy, *A Consumer's Guide: Get Your Power from the Sun,* DOE/GO-102003-1844, December 2003.

[9] International Renewable Energy Agency, *Concentrating Solar Power, Renewable Energy Technologies: Cost Analysis Series,* Volume 1: Power Sector, Issue 2/5, June 2012.

[10] International Energy Agency, *Technology Roadmap: Concentrating Solar Power,* 2010.

[11] Hague, R.D., and P. Asmus, *Concentrated Solar Power,* Research Report, 4Q 2011, Pike Research, 2011.

[12] Solar Energy Industries Association, *U.S. Solar Market Insight Report,* Q2 2012, 2012.

[13] International Energy Agency, *Technology Roadmap: Solar Photovoltaic Energy,* 2010.

[14] International Renewable Energy Agency, *Wind Power, Renewable Energy Technologies: Cost Analysis Series,* Volume 1: Power Sector, Issue 5/5, June 2012.

[15] International Energy Agency, *IEA Wind 2011 Annual Report,* July 2012.

[16] American Wind Energy Association, *2011 U.S. Small Wind Turbine Market Report (Year Ending 2011),* 2011.

[17] American Wind Energy Association, *FAQ for Small Wind Systems,* Fact Sheet.

[18] Gauntlett, D., and P. Asmus, *Small Wind Power, Research Report,* 3Q 2011, Pike Research, 2011.

[19] American Wind Energy Association, *Offshore Wind: America's New Energy Opportunity,* Fact Sheet.

[20] International Energy Agency-Renewable Energy Technology Deployment (IEA-RETD), *Offshore Renewable Energy: Accelerating the Deployment of Offshore Wind, Tidal and Wave Technologies,* New York, NY: Earthscan, 2012.

[21] Schwartz, M., et al., *Assessment of Offshore Wind Energy Resources for the United States,* Golden, CO: National Renewable Energy Laboratory, 2010.

[22] U.S. Department of Energy, *20% Wind Energy by 2030: Increasing Wind Energy's Contribution to the U.S. Electricity Supply,* 2008.

[23] Schwartz, M., et al., *Assessment of Offshore Wind Energy Resources for the United States,* NREL/TP-500-45889, Golden, CO: National Renewable Energy Laboratory, June 2010.

[24] American Wind Energy Association, *Utility Scale Wind Energy and Sound,* Fact Sheet.

[25] BTM Consult, *International Wind Energy Development: World Market Update 2011 (Forecast 2012-2016),* March 2012.

[26] American Wind Energy Association, *AWEA U.S. Wind Industry Third Quarter 2012 Market Report,* October 17, 2012.

[27] World Wind Energy Association, *2012 Half-year Report,* October 2012.

[28] Navigant Consulting, Inc., *Impact of the Production Tax Credit on the U.S. Wind Market, Report prepared for the American Wind Energy Association,* December 12, 2011.

[29] Glassley, W. E., *Geothermal Energy (Renewable Energy and the Environment Series),* Boca Raton, FL: CRC Press, 2010.

[30] Website: Calpine Corporation, *The Geysers: Welcome to the Geysers,* Available at: http://www.geysers.com/.

[31] Geothermal Energy Association, *Annual U.S. Geothermal Power Production and Development Report,* April 2012.

[32] Website: U.S. Department of Energy, Geothermal Technologies Program, Electricity Generation, Available at: http://www1.eere.energy.gov/geothermal/powerplants.html.

[33] Blodgett, L., "Low-Temperature Geothermal Activity is Taking Off," *Renewable Energy World,* December 1, 2010.

[34] Lawrence, M., and C. Wheelock, *Geothermal Power, Research Report,* 1Q 2011, Pike Research, 2011.

[35] Lawrence, M., and C. Wheelock, *Geothermal Heat Pumps and Direct Use, Research Report,* 3Q 2011, Pike Research, 2011.

[36] Website: U.S. Department of Energy, Geothermal Technologies Program, Low-temperature and Co-produced Geothermal Resources, Available at: http://www1.eere.energy.gov/geothermal/low_temperature_resources.html.

[37] Website: U.S. Department of Energy, Geothermal Technologies Program, Enhanced Geothermal Systems, Available at: http://www1.eere.energy.gov/geothermal/enhanced_geothermal_systems.html.

[38] Entingh and Vimmerstedt, 2005.

[39] Lawrence, M., and C. Wheelock, *Biopower Markets and Technologies, Research Report,* 1Q 2012, Pike Research, 2012.

[40] Boundy, B., et al., *Biomass Energy Data Book: Edition 4, ORNL/TM-2011/446,* Oak Ridge, TN: Oak Ridge National Laboratory, September 2011.

[41] International Renewable Energy Agency, B*iomass for Power Generation, Renewable Energy Technologies: Cost Analysis Series,* Volume 1: Power Sector, Issue 1/5, June 2012.

[42] International Energy Agency, *Technology Roadmap: Bioenergy for Heat and Power,* 2012.

[43] Bracmort, K., *Is Biopower Carbon Neutral?,* CRS Report for Congress, July 6, 2011.

[44] Organisation for Economic Co-operationand Development, *Bioheat, Biopower, and Biogas: Developments and Implications for Agriculture,* 2010.

[45] Lawrence, M., and K-A. Adamson, *Renewable Biogas, Research Report,* 2Q 2012, Pike Research, 2012.

[46] Schubert, R., et al., *Future Bioenergy and Sustainable Land Use,* London, UK: Earthscan 2010.

[47] U.S. Department of Energy-Biomass Program, *Integrated Biorefineries: Biofuels, Biopower, and Bioproducts,* Program Fact Sheet, July 2012.

[48] Lawrence, M., and K-A. Adamson, *Waste-to-Energy Technology Markets,* Research Report, 1Q 2012, Pike Research, 2012.

[49] International Energy Agency, Technology Roadmap: Hydropower, 2012.

[50] International Renewable Energy Agency, *Waterpoer, Renewable Energy Technologies: Cost Analysis Series,* Volume 1: Power Sector, Issue 3/5, June 2012.

[51] Website: Homepower, Microhydro Power Basics, Available at: https://homepower.com/articles/what-microhydro-power.

[52] Natural Resources Canada, *Micro-Hydropower Systems: A Buyer's Guide,* 2004.

[53] Gauntlett, D., and P. Asmus, *Hydrokinetic and Ocean Energy, Research Report,* 1Q 2012, Pike Research, 2012.

[54] U.S. Department of Energy-Wind and Water Program, *Water Power for a Clean Energy Future,* GPO DOE /EE-0661, June 2012.

[55] Carbon Trust, *Marine Energy Briefing,* July 27, 2012.

4

Alternative Fuels

4.1 Key Facts, Figures, and Trends

Some of the key trends in the United States and worldwide relating to alternative fuels are highlighted in the following sections.

4.1.1 United States

As of May 2012, there were a total of 17,196 alternative fueling stations (types and specific numbers shown in Table 4.1). Electricity charging stations dominate, followed by liquefied petroleum gas (LPG), and ethanol (blended in gasoline at 85 percent; E85) [1].

The majority of CNG, LNG, electricity, and hydrogen stations are located in California; E85 stations are located in Minnesota; Biodiesel stations are located in North Carolina; LPG stations are located in Texas [1].

The top five ethanol producing states are Iowa, Nebraska, Illinois, Minnesota, and Indiana [2]. These states are located in what is called the "corn belt" of the United States, and reflect that though there is growing interest in cellulosic-based ethanol, current production in the United States remains corn-based. Minnesota requires an E10 in all its gasoline, and E85 is available at approximately 360 fueling stations statewide [1].

The United States exported a record one billion gallons ethanol in 2011 (representing an increase of 600 million gallons over 2010) mainly to Brazil, Europe, and Canada [3].

In terms of alternative fuel station deployment between 1992 and 2012 CNG and LPG were the dominant fuels, while other alternatives started entering the mix later. E85 stations have steadily increased during the past seven

Table 4.1
Alternative Fueling Stations in the U.S. (as of May 2012) [1]

B20	CNG	E85	Electricity	Hydrogen	LNG	LPG
618	1,014	2,499	10,296	56	52	2,661

years. Electric charging points grew dramatically in 2010, with a greater increase between 2011 and 2012. As of May 2012, electric charging points made-up approximately 65 percent of the stations, while E85 stations were 15 percent of the total [1].

Since 1995, vehicles using E85 have rapidly increased, reaching approximately 66 percent of the total in 2010 [1].

4.1.2 Worldwide

The top three countries for installed base of natural gas stations and vehicles are Iran, Pakistan, and Argentina. Iran leads with NGVs (2.86 million) and has 1,800 stations; Pakistan has 2.85 million NGVs and leads with stations (3,330); and Argentina has 2.08 million NGVs and 1,913 stations [4].

LPG is the third most used transportation fuel worldwide, with great popularity in countries like Japan, Canada, Italy, and Austria [5].

Both the global ethanol supply and demand in 2011 was dominated by the United States, followed by Brazil [6].

It has been projected that corn, sugarcane, rapeseed, and soy will be the major feedstocks of choice for biofuels over the next decade. However, advanced production technologies and nonfood-based feedstocks are viewed to have the potential to create a larger biofuel market worldwide. At present, the three major biofuel markets belong to the United States, Brazil, and the European Union [7].

Global EV and PHEV sales are expected to rise steadily (especially after 2020) and reach close to 120 million vehicles by 2050 [8].

Global algae-based biofuels markets are projected to grow at a CAGR of 72 percent, reaching 61 million gallons of production per year by 2020, with a market value of $1.3 billion. Latin America and the Asia Pacific regions are projected to grow significantly in the long-term, due to abundant land and water resources, as well as access to cheap labor [9].

It has been forecast that the worldwide demand for hydrogen will be more than 400 million kg annually, representing an 88 percent CAGR. Initially, the global demand for hydrogen is expected to be lower and dominated by forklifts and uninterruptible power sources (UPS). Demand is then forecast to rise significantly, prompted by the expected introduction of commercial hydrogen fuel cell LDVs around 2015 [Pike_H2 infrastructure]. FCEV commercialization is expected to be in the hundreds of units in 2012 and in the thousands of units by 2015, led mainly by Germany, Japan, and Korea.

4.2 Basics

Alternative fuels may be produced from fossil or renewable sources, allowing for the availability of a wider range of fuels. However, the differing characteristics and production pathway of each fuel results in diverse prices, benefits, development pathways, challenges, and opportunities. Alternative fuels may also be categorized as gaseous fuels (compressed natural gas, liquefied natural gas, and liquefied petroleum gas), alcohols (methanol and ethanol), biodiesel, hydrogen, and electricity. The gaseous fuels are mainly fossil-based and thus have environmental impacts (greenhouse gas and other air pollutant emissions) and price volatility (as they are dependent on volatile-priced fossils fuels). The main characteristics of various alternative fuels compared to gasoline are provided in Table 4.2.

4.2.1 Liquefied Petroleum Gas

Liquefied petroleum gas (LPG) is a byproduct of petroleum refining or natural gas processing and is commonly used in internal combustion engine vehicles (ICEVs). It is present in natural gas and oil deposits, and it must be purified or refined in order for natural gas and petroleum to produce saleable products. Generally, 3 to 40 percent of a barrel of oil can be refined into LPG [5]. At normal temperatures and pressures, the hydrocarbons present in LPG (mainly propane) are in gas form. Stored under pressure, propane turns into a liquid. The typical spark-ignition engine vehicles that utilize LPG can make use of either vapor- or liquid-injection.

Besides its use as an alternative fuel in transportation, LPG is also used in applications such as heating (home and water), cooking, powering farm and industrial equipment, drying corn, and as a raw material for making plastics and other compounds. LPG is mainly composed of propane (C_3H_8). To be sold as an alternative fuel, LPG is also required to include other gases the composition of which is 90 percent propane, five percent propylene, and 5 percent other gases (mainly butane, C_4H_{10}) [1, 5]. A higher percentage of *propane* is used in the winter, while a higher percentage of *butane* is used in the summer. As propane is the major component of LPG, this alternative fuel is commonly referred to as propane.

A gallon of LPG has 73 percent of the energy of one gallon of gasoline [1]. LPG has several characteristics that make it a favorable alternative fuel. The sulfur content in LPG is lower than that of gasoline, providing for lower sulfur emissions when used in an ICEV. Overall emissions of other pollutants—carbon

Table 4.2
Comparison of the Properties of Alternative Fuels with Gasoline [1]

	Gasoline	Biodiesel	CNG	Electricity	Ethanol	LNG	LPG	Methanol
Main Fuel Source	Crude Oil	Fats and oils from sources such as soy beans, waste cooking oil, animal fats, and rapeseed	Underground reserves	Coal, nuclear, natural gas, hydro, wind, and solar	Corn, grains, or agricultural waste	Underground reserves	By-product of petroleum refining or natural gas processing	Natural gas, coal, or woody biomass
Energy Content (lower heating value)	116,090 Btu/gal (g)	119,550 Btu/gal for B100 (g)	20,268 Btu/lb (g)	3,414 Btu/kWh	76,330 Btu/gal for E100 (g)	74,720 Btu/gal (g)	84,950 Btu/gal (g)	57,250 Btu/gal (g)
Energy Contained—as Compared to One Gallon of Gasoline	100%	B100 has 103% of the energy in one gal of gasoline or 93% of the energy of gal of diesel (for B20: 109% and 99%, respectively)	5.66 lbs or 126.67 cu. ft. of CNG has 100% of the energy of one gal of gasoline	33.70 kWh has 100% of the energy of one gal of gasoline	1 gal of ethanol has 77% of the energy of one gal of gasoline	1 gal of LNG has 64% of the energy of one gal of gasoline	1 gal of LPG has 73% of the energy of one gal of gasoline	1 gal of methanol has 49% of the energy of one gal of gasoline
Physical State	Liquid	Liquid	Compressed Gas	Electricity	Liquid	Cryogenic Liquid	Pressurized Liquid	Liquid

monoxide, nitrogen oxide, particulate matter, and greenhouse gases (GHGs)—are also lower for propane-utilizing vehicles, compared to gasoline powered vehicles. Other advantages of the fuel include its high octane number, narrow flammability range, ability to be transported as a liquid, and its relatively lower cost (on a per gallon basis).

The high octane rating of LPG (105, compared to 84–93 for gasoline [1]) allows for increased vehicle fuel efficiencies, as well as improved engine life, as much as two times that of gasoline engines [10]. While the higher octane rating provides benefits, its lower energy content (lower Btu per gallon) compared to gasoline means that more of the fuel needs to be used to travel the same distance, resulting in lower fuel economy.

In terms of infrastructure, the needs of an LPG fueling infrastructure are similar to a gasoline infrastructure. LPG is transported to the refueling site via trucks and stored above-ground in tanks. Dispensers of LPG can be located along with dispenser of gasoline, diesel, and other alternative fuels at conventional refueling sites.

4.2.2 Natural Gas

Natural gas may be used in vehicles in compressed gas or liquid form as compressed natural gas (CNG) or liquefied natural gas (LNG). While CNG is stored in tanks at pressures of 3,000 to 3,600 pounds per square inch (psi), LNG is cooled down to temperatures as low as −260°F and stored in insulated tanks [11]. More natural gas can be stored in tanks in the liquefied form, as the liquid is denser than gas, and this liquid form provides for the longer driving range needs of trucks. Use of natural gas in vehicles can be in dedicated natural gas vehicles, or dual-fuel vehicles which are able to operate on both natural gas and gasoline/diesel. While use in dual-fuel vehicles provides flexibility in fuel use, the requirement of two different fueling systems decreases the vehicle's passenger and cargo space.

As natural gas contains lower amounts of carbon, compared to gasoline or diesel vehicles, NGVs emit lower amounts of various air pollutants, including carbon monoxide, nonmethane hydrocarbons, nitrogen oxides, and carbon dioxide. Use of natural gas in vehicles results in 60 to 90 percent decrease in smog-forming pollutants, and 30 to 40 percent less GHGs.

In comparison to conventional fuels, natural gas is relatively safe, as it has higher explosive limits and is lighter than air (thus can disperse rapidly if a leak occurs). For a continuous flame spread to occur, the natural gas-air mixture needs to be a minimum of five percent by volume, compared to two percent for propane and one percent for gasoline [12].

CNG and LNG fueling stations have different configurations and requirements. While CNG stations have more complex equipment requirements, LNG stations require more safety precautions.

CNG stations can either be *fast-fill* or *slow-fill*, with the available storage capacity and compressor size determining the amount of fuel to be dispensed and the time required. Fast-fill is typically used, along-side gasoline, in retail stations to meet the refueling needs of light-duty vehicles. Natural gas is delivered to these stations, under low pressure, via utility lines and is compressed onsite to higher pressures, after which it is stored in a series of high-pressure storage vessels. This allows for the quick dispensing, approximately three minutes for automobiles and six minutes for buses [12].

Slow-fill stations for CNG are better suited for fleet vehicles which refuel at large central locations at night and are able to handle longer refueling times. Similar to fast-fill stations, natural gas is also delivered to the stations via utility lines and under low pressure. However, the difference here is that the compression of the gas to higher pressures is done by means of larger compressors and the fuel is dispensed to the vehicles directly from the compressors. Thus, large storage tanks are not required, but small buffer tanks are used. The refueling time could range from a couple minutes to a number of hours, depending on the compressor size, amount of buffer storage, and the size and number of vehicles to be fueled. As the time of refueling is controlled and done mainly at night, the station may make use of lower-priced off-peak electricity, which reduces the costs of running the larger compressors required for these types of refueling sites. As the refueling needs of various fleets (buses, garbage trucks, etc.) differ, slow-fill CNG stations need to be designed according to and specifications of these fleets.

LNG stations dispense liquid fuel, making them similar in structure to gasoline refueling stations. However, as the fuel is at very cold temperatures, protective measures and clothing are required when dispensing. The dispensing structure of LNG may vary from a mobile station (a tanker truck with dispensing onboard) to custom stations with all the required equipment and tailored to meet a fleet's needs.

4.2.3 Methanol

Methanol is an alcohol fuel, commonly known as wood alcohol, and can be produced from a range of carbon-containing sources, such as natural gas, coal, and also renewable resources such as wood and landfill gas. It has higher energy content per volume when compared to other alternative fuels such as CNG and LPG [13]. It also has the benefits of a high octane rating and less flammability. Methanol may be used directly, blended with gasoline or diesel, or converted to dimethyl ether for use. If blended, percentages ranging from 5 to 85 percent are typically used [13].

Use of methanol directly in passenger vehicles would require substantial engine modifications and changes in the lubrication system. Additionally, cold-start problems may occur in this case, as methanol does not contain a sufficient amount of highly volatile compounds (which provide vapor to the engine for ignition in cold climates). Thus, volatile compounds like butane and propane would need to be added to improve engine performance in cold climates. Methanol may also be blended with diesel (at no more than 30 percent by weight for optimum performance), but methanol's corrosive characteristic necessitates special considerations in engine design [13].

A derivative of methanol, dimethyl ether, is another option for this alternative fuel's use, and is produced by dehydrating methanol. This variation is known as a ultra-clean fuel which works well with diesel engines (along with other applications such as power generation).

Although methanol is a toxic substance, several studies have concluded that its use as a transportation fuel does not pose major concerns. While the production, transport, and use of methanol may cause it to be released into the environment, as with any fuel, necessary precautions need to be taken to prevent contamination. If contamination does occur, methanol is biodegradable in the environment [13]. A study by the U.S. Department of Energy (DOE) has concluded that methanol is less toxic to humans than gasoline [14]. Additionally, the U.S. Environmental Protection Agency (EPA) has rated methanol as low as seven on a scale of one to 100 [15]. While methanol is flammable, it has several characteristics that reduce the effects of this danger. It is harder to ignite when compared to gasoline, and if ignited, burns cooler, at a 60 percent slower rate, and with less smoke [13].

4.2.4 Ethanol

Ethanol is a renewable alcohol fuel that may be produced from a variety of biomass resources such as starch- and sugar-based feedstocks (e.g., corn, sugar cane), or cellulosic feedstocks (e.g., dedicated energy crops, wood residues, crop residues, industrial waste). Sugar-based feedstocks are easy to process, allowing for this production pathway to be more affordable and popular. In the United States, starch-based corn is currently the dominant feedstock of choice for ethanol production. Cellulosic feedstocks are nonfood based feedstocks, providing the benefit of not creating competition for food choices (like the starch-based feedstocks do), but due to their complex chemical natures, it is harder to separate the sugars from these resources to be able to convert to ethanol. If cellulosic feedstocks originate form waste materials, it provides an added advantage of managing wastes. If specific energy crops are used, they often may be grown on marginal land. However, achieving ethanol's potential requires transitioning from corn to cellulosic perennial grasses, crop residues, and wood.

Ethanol is nontoxic, less flammable than gasoline, water soluble and biodegradable. While the combustion of ethanol results in the emission of carbon dioxide, this is balanced by the carbon dioxide captured when plants are grown for use as feedstock for ethanol. Thus, net zero or lower carbon emissions are possible.

While the use of ethanol in internal combustion engines was considered back in 1897, use of this alternative fuel started gaining more interest during the fuel crises of World War I and II. In the United States, during the energy crisis of the 1970s, use of ethanol advanced as regulations and standards were introduced regarding vehicle emissions. Oxygenated fuels were required to provide for the reductions in emissions, corresponding to approximately 7.5 percent ethanol, by volume, to be used in gasoline vehicles. Several other countries also started using ethanol in gasoline blends [16].

Ethanol may be blended in percentages ranging from 10 percent (called E10) to 85 percent (called E85). In the United States, ethanol may be found as an E10 blend (10 percent ethanol) in more than 95 percent of the all gasoline sold at retailers, and represents nearly 10 percent of the nation's gasoline supply [2]. It is blended up to 25% in Brazil. As it has a higher octane number than gasoline, it allows for gasoline to attain its standard 87 octane requirement and also helps prevent problems in vehicle engines during cold climates, by acting as an anti-freeze. Ethanol's energy content (76,330 Btu/gal for E100, on a LHV basis) is lower than gasoline's energy content (116,090 Btu/gal, on a LHV basis). While this results in lower fuel efficiency for vehicles using ethanol, the lower temperature burning of ethanol is gentler on a vehicle's engine, providing for less wear-and-tear and thus longer engine life.

In the United States, ethanol-blended fuels are approved under the warranties of all auto manufacturers that market vehicles in the country and some even recommend ethanol use for its clean burning benefits. Even many manufacturers of power equipment, motorcycles, snowmobiles and outboard motors permit the use of ethanol blends in their products [17].

The main process of ethanol production involves turning biomass materials into fermentable sugars. Grain crops such as corn or wheat also require the processes of milling and starch separation. While sugar-based feedstocks are simpler to process, cellulosic feedstocks are more complex, as the sugars in these materials are locked in complex carbohydrates, and thus require an additional process called *saccharization (or enzymatic hydrolysis).* The main ethanol production steps are feedstock harvesting, conversion of feedstock to sugar, and conversion of sugar to alcohol (fermentation and distillation of alcohol), as seen in Figure 4.1. These processes require energy: grain crop conversions generally use fossil fuels for this energy, while sugar crop and cellulosic material conversions

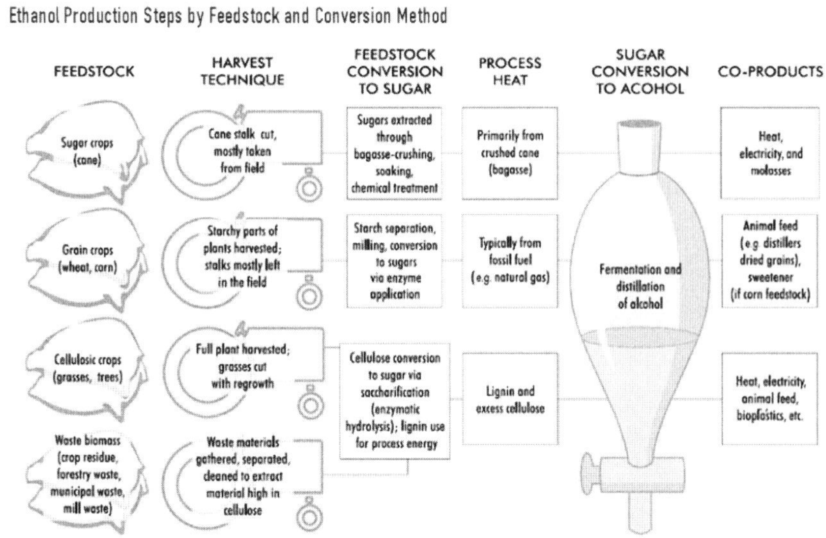

Figure 4.1 Ethanol production process based on feedstock type [18].

have residual biomass materials that can be used for process heat. If grain crops (such as corn or wheat) are used as feedstock, two methods are initially used to treat these grains, dry milling and wet milling. In dry milling, the entire starchy grain is first ground into flour and then processed without separating out the various component parts of the grain. On the other hand, in wet milling, the grain is soaked in water and dilute sulfurous acid for duration of 24 to 48 hours, which facilitates the separation of the grain into its component parts [18].

The range of feedstocks that may be used in cellulosic ethanol production include:

- *Municipal and Other Wastes:* Organic wastes such as municipal solid waste, waste from food processing facilities, paper pulp, and paper products.
- *Agricultural Residue:* Residues from crop processing, such as wheat straw and corn stalks, leaves, and husks.
- *Forestry Residue:* Residues from logging and milling, such as wood chips, sawdust, and pulping liquor.
- *Energy Crops:* Grasses (fast-growing grasses like switchgrass); trees (fast-growing trees like willow and poplar).

While ethanol is a popular and developing fuel, there have been many misconceptions regarding its impacts. The Renewable Fuels Association

[3] has compiled fact-based explanations to several of the most common misconceptions:

Myth: Ethanol uses too much corn.

In the United States, only 26 percent of the country's corn crop is utilized for ethanol on a net basis, and the U.S. corn harvest in 2011 was fourth largest in history, while the global supplies have reached the largest levels in history. The production process of ethanol consumes only two-thirds of a bushel of corn and the remaining amount is used as feed for livestock.

Myth: Ethanol production raises the price of corn and food.

An absolute, direct connection cannot be made, as the price of food is affected by a variety of factors, including oil prices, speculation in commodities, weather, and financial policies.

Myth: Ethanol uses more energy than it yields.

Considered from a *life-cycle* perspective, the production of ethanol yields 1.7 to 2.3 units of energy for every unit of energy used throughout the process. Additionally, as a comparison, a gallon of ethanol involves 19 times less crude oil to produce than a gallon of gasoline.

Myth: Ethanol production uses too much water.

Advancements of ethanol production technology have allowed the reduction of water requirements up to 47 percent (since 2001). In comparison, the processing of marginal oil resources requires up to eight barrels of water for each barrel of oil produced.

4.2.5 Biodiesel

Biodiesel can be manufactured from a variety of resources, such as vegetable oils (new or used, such as soybean, rapeseed, cottonseed, or palm oil), animal fats, and recycled restaurant grease, and may be blended up to 100 percent with petroleum diesel in most engines, though some minor modifications may be required. In the United States, soybean is the primary feedstock for biodiesel production; whereas in Europe, the main feedstock choice is rapeseed oil. The use of used/recycled oils serves the dual purpose of handling waste and generating an alternative, more environmentally-friendly fuel, and prevents the issues surrounding utilizing food crops. More importantly, though, is the fact that similar to the need for ethanol production to move towards using cellulosic materials, biodiesel must move more towards utilizing algae, to enable a more sustainable development path for this alternative fuel.

If biodiesel is blended up to five percent (B5) with petroleum diesel, it will remain in the confines of the ASTM D975 diesel fuel specification, and thus may be used as if it was conventional diesel without any needs for alterations to vehicle engines. In some higher concentrations (blends of 6 to 20 percent, B6–B20) use of biodiesel requires only minor modifications to engines. While use of pure or 100 percent biodiesel (B100) is possible, more precautions need to be taken to do so (special handling procedures, and more substantial engine modifications). In the U.S., B20 is the most popular blend, and is also the minimum blend specification of the alternative fuel requirements of the Energy Policy Act of 1992 (EPAct 1992) [19].

Two main challenges of biodiesel production are (1) meeting the high ASTM standards for diesel production and (2) converting potentially hazardous by-products glycerol and methanol into marketable quality product. Biodiesel also touts beneficial safety aspects, such as biodegradability, less toxicity (10 times less than table salt), and a high flashpoint (110°C, compared to 45°C–55°C for conventional diesel) [20].

The use of biodiesel offers important reductions in air emissions. As biodiesel contains 11 percent oxygen by weight, this oxygen content allows fuller combustion of the fuel in engines, thus reducing the emissions of incomplete combustion, such as particulate matter (PM), hydrocarbons, and carbon monoxide (CO). Moreover, it has been demonstrated that the reductions in these pollutants are not dependent on the type of feedstock used for biodiesel production [19]. It has also been shown that over the life cycle, the production and use of biodiesel generates up to 80 percent reductions in GHGs, and close to 100 percent reductions in sulfur dioxide, compared to petroleum diesel. While the GHG savings are also substantial, the elimination of sulfur dioxide emissions is also a key benefit, especially in regions where local air pollution is of concern. Depending on the engine type and testing procedure, use of biodiesel has been shown to either increase or decrease in nitrogen oxide emissions [8x]. In 2006, the United States lowered the permitted level of sulfur in diesel fuel to 15 ppm, a significant drop from the 500 ppm allowed previously. This requirement supports the integration of biodiesel blends because the sulfur content of biodiesel is extremely low.

The energy content of biodiesel is approximately seven percent less than conventional No. 2 diesel (119,550 Btu/gal for biodiesel, versus 127,960 Btu/gal for No. 2 disel, both on LHV basis) [1]. While the impact of this difference to fuel economy will be more apparent in B100, for lower blends like B20, differences only approximately one to two percent have been seen, and blends as low as B5 demonstrate no significant difference [19].

Similar to conventional diesel, biodiesel is apt to freeze or gel under typical winter temperatures, becoming too thick to pump. However, one benefit of biodiesel over diesel is that biodiesel's freeze point is 20°F to 30°F higher than that of petroleum diesel, thus extending the temperature ranges for successful operation of conventional diesel vehicles. Biodiesel is more susceptible to microbial degradation, which is an advantage in the case of spills (allows for rapid biodegradation in the environment, causing less damage), but a disadvantage for vehicle use, as it may cause the vehicle to stall. However, proper storage tank maintenance helps overcome this problem [19].

In terms of producing biodiesel, vegetable oils (e.g., soybean oil) and animal fats are made into methyl esters (i.e., fatty acid methyl esters [FAME]) by a process called *transesterification.* In the process, vegetable oils are mixed with an alcohol (usually methanol) and a catalyst to produce FAME and glycerin by-product. Excess methanol is then removed by distillation and recycled in the process. The resulting FAME is a liquid with properties very similar to those of diesel fuel, enabling it to be used as fuel in diesel engines [21]. In general, 100 pounds of oil or fat react with 10 pounds of alcohol to form 100 pounds of biodiesel and 10 pounds of glycerin [19].

The chemical reactions will typically progress either at a very slow rate or not at all. So, heat is used to help the reaction proceed more quickly. Assuming an alkali catalyst is used, the reaction can be completed in approximately 90 minutes at 60°C (140°F) under atmospheric pressure since methanol evaporates at 65°C under atmospheric pressure [21]. Variations of temperature, pressure, catalyst materials, and alcohols can be used in the transesterification process, but they usually add to overall cost of the operation.

In addition to traditional transesterification, supercritical alcohols (e.g., methanol, ethanol) are currently being researched to simultaneously act as mediums in oil extraction as well as catalysts in the transesterification process. Combining oil extraction and transesterification into one step has the potential to significantly improve production efficiency. To date, this process has only been demonstrated with vegetable oils but can be applied to other feedstocks, including algae. As with traditional transesterification, the supercritical fluid processes require certain levels of heat and electricity [22].

In the United States, soybean oil is the vegetable oil of preference to produce biodiesel fuel. This is likely to be the case for a number of years until the practice of using a vegetable oil as a stream into petroleum refineries is adopted in the United States as it has been in some European countries. The major (and minor) soybean growing areas within the continental United States (Minnesota, Iowa, Illinois, Indiana, and Ohio) are considered the top soybean producing states, accounting for nearly 60 percent of the entire national soybean crop.

4.2.6 Algae-Based Biodiesel

Algal biomass offers a promising pathway for the production of biodiesel. Algae contain substantial amounts of lipids that can be converted into this renewable alternative fuel. Currently, an established commercial market for algae-based biodiesel does not exist. While developments are ongoing, realizing the full potential of this alternative will necessitate improvements in processing technologies, as well as related fundamental biology.

Algae are simple photosynthetic microorganisms that utilize water and carbon dioxide to generate biomass in a more efficient way than land-based plants. Due to their simple structure, algae are able to convert roughly six percent of the incoming solar radiation into biomass, as opposed to the 3.5 to 4 percent solar conversion efficiencies of sugarcane (which is considered one of the most productive crops). A variety of algal strains have been tested in the laboratory, demonstrating that algae are able to produce over 50 percent of their weight as lipids (varying in quantity and quality, depending on growth conditions). If nutrients such as nitrogen and phosphorus are constrained, algae are able to generate higher percentages of lipids. This comes at the expense of lower biomass yields, but still does not take away from the possibility of algae producing much higher levels of oil than conventional oilseed crops. Algae are typically cultivated in nonarable land, brackish water, saline water, or wastewater, thus making use of otherwise nonvaluable resources and not creating competition for food or land resources. Algae-based biodiesel production also has a unique place in the future carbon-constrained world in that cultivation ponds to grow algae could be co-located with fossil-based power plants, in an integrated biorefinery, and thus reduce the emissions of GHGs (by using the GHGs in their growth process) [23].

The cultivation of algae may occur either in open ponds or closed tubes (photobioreactors). While both methods have their unique benefits and downfalls, closed tube systems are more expensive to construct, as their engineering has not yet been optimized for commercial use. Larger-scale systems need to be built and demonstrated over several years to gain further insight and discover means to improve costs to be commerciallyviable. Another factor in the successful commercial potential of algal biodiesel relates to algal growth. There is a need for the development of algal strains that are able to deliver rapid biomass development with high lipid content and minimum competing strains. While the presence of contaminating or competing algal strains affects both the productivity and the economics of the process, one unique benefit is that algal ponds are able to resume production in a couple of days once they have been reinoculated, as opposed to the whole growing cycle of land-based crops being affected when damage to growth occurs. Additionally, algae are able to successfully grow in a wide range of climatic conditions, with the exception of a few strains which have more limitations. If adequate solar radiation is available, the theoretical maximum yield for algae is 100 g/m2/day. However, current reported highest yield levels are in the range of 15–30 $g/m^2/day$ and can even be lower if factors such as the area required for supporting activities (water supply, treatment, disposal, etc.) are included in the calculations as well. Closed systems may provide higher yields than open pond systems, due to the higher surface-to-volume ratio [23].

Oil yields provided by algae (liter/hectare) can be up to 20 times more than yields from soybeans, thus requiring a smaller area for production. Once algae are cultivated, they need to be harvested for biodiesel production. However, commercially acceptable low cost harvesting has not been developed, and more engineering research efforts are needed to establish cost-effective algal harvesting methods. While algal biodiesel has benefits and promises, it also faces other challenges, such as the availability of land with the required characteristics (soil type, slope, elevation, incident radiation, temperature, etc.), as well as the proximity of this land to adequate water and carbon dioxide, and end-use markets and transportation infrastructure. More data needs to be accumulated regarding the geographic potential of algal biodiesel production, before a more complete understanding is reached and claims can be made on the potential of this pathway to replace some or all of petroleum-based transportation fuels [23].

In addition to the extraction of lipids from algae, two of the organism's other core elements, carbohydrates and proteins, may also be extracted. This allows for the production of various chemical intermediates that may be converted into different fuels, such as alcohols, diesel, methane, and hydrogen, as shown in Figure 4.2 [23].

Nitrogen and phosphorus are key elements that aid the growth of algae. If algae are cultivated using wastewater, nitrogen and phosphorus in the wastewater would feed these organisms, which would mean that there would be a reduced need for the consumption of fertilizers during the process, reducing the

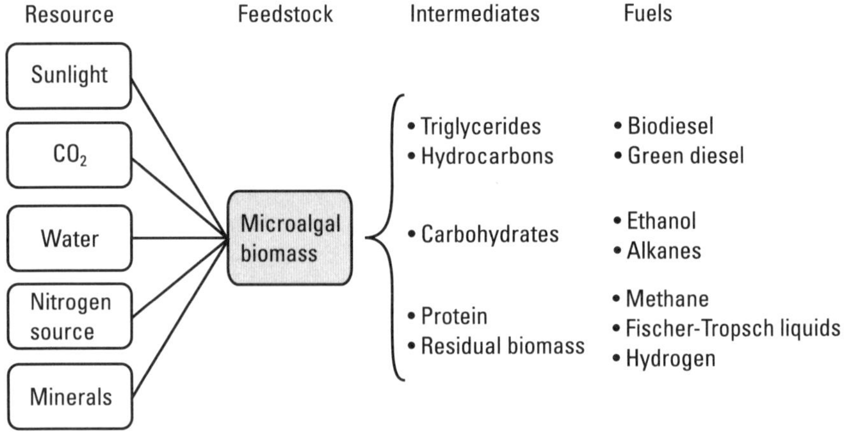

Figure 4.2 Using algae to create a variety of fuels [21].

environmental and energy impacts of fertilizer production. The consumption of water is also an important consideration in algae-based biodiesel production. The fundamental photosynthesis reactions that the algae undergo consume water. Additionally, especially if open ponds are being used to cultivate algae, the water of the algae culture needs to be replenished frequently (to avoid bacteria and toxics build-up), while the existing water in the ponds also evaporates. The location of the cultivation pond, and its climate, is an influential factor. More evaporation occurs in warmer climates, whereas the evaporation could be tolerated and somewhat offset in climates where there is adequate rainfall. If algae were to be cultivated at larger scales, to meet a larger portion of transportation fuel demand, then larger land areas would be required for cultivation. While the land typically used for algae cultivation is marginal land, covering a large portion of any land with just one species will lead to loss of biodiversity and more vulnerability of the ecosystem to diseases [24].

A 2011 report [24] from the International Energy Agency (IEA) Advanced Motor Fuels Implementing agreement evaluated the state of algal biofuels technology and opportunities. The production costs of algal biofuels utilizing current technology were found to range between approximately $8/gal and $40/gal, much higher than the day's petroleum fuel production costs. A survey of various market forecast summarized in this report also showed that future algal biofuel prices would range between $1.40/gal and $3.00/gal, which would be competitive with petroleum-based fuels. The IEA report also quoted a projection by SBI Energy that estimated the total algal biofuels market to reach $1.6 billion by 2015, representing 43 percent annual growth between 2010 and 2015.

4.2.7 Hydrogen

In 2010, worldwide hydrogen production totaled 31.3 million MT, with U.S. hydrogen production composing 20.1 million MT. The majority of hydrogen is captive (i.e. produced and consumed on-site by hydrogen end-user) and has historically been used primarily in ammonia production and in oil refineries. In the United States, currently, over 95 percent of hydrogen is produced from natural gas. Worldwide, close to 96 percent of all hydrogen is derived from fossil fuels, with 49 percent coming from natural gas, 29 percent from liquid hydrocarbons, 18 percent from coal, and 4 percent from electrolysis and other byproduct sources. However, the demand base is widening with significant expansion into the hydrogenation of edible fats and oils, metallurgy/furnace annealing, and the electronics markets (see Table 4.3). Today, hydrogen's use as an *energy carrier* is developing in order to fuel cars and buses, to produce prime and back-up electricity, to replace batteries in forklifts, and to power consumer electronics.

Table 4.3
Current Applications of Hydrogen in Industry [25]

Food	To hydrogenate liquid oils (e.g., soybean, fish, cottonseed and corn), converting them to semisolid materials such as shortenings, margarine, and peanut butter. Hydrogen is also an authorized food additive (E 949) that allows food package leak testing among other anti-oxidizing properties.
Chemical Processing	Primarily to manufacture ammonia (i.e., fertilizer market) and methanol, but also to hydrogenate non-edible oils for soaps, insulation, plastics, ointments, and other specialty chemicals.
Metal Production and Fabrication	To serve as a protective atmosphere in high-temperature operations such as stainless steel manufacturing. Commonly mixed with argon for welding austenitic stainless. Also, used to support plasma welding and cutting operations. It is used as shielding gas in welding methods.
Pharmaceuticals	To produce sorbitol used in cosmetics, adhesives, surfactants, and vitamins A and C.
Aerospace	To fuel spacecraft, but also to power life-support systems and computers, yielding drinkable water as a by-product.
Electronics	To create a specially controlled atmosphere in the production of semiconductor circuits.
Petroleum Recovery and Refinery	To enhance performance of petroleum products by removing organic sulfur from crude oil, as well as to convert heavy crude to lighter, more easily refined, and more marketable products. Hydrogen's use in reformulated gas products helps refiners meet Clean Air Act requirements. Methods used for these refining processes include hydrodealkylation, hydrodesulfurization, and hydrocracking.
Power Generation	To serve as a heat transfer medium for cooling high-speed turbine generators. Also, used to react with oxygen in the cooling water system of boiling water nuclear reactors to suppress intergranular stress corrosion cracking in the cooling system.
Glass Making	Flat "float glass" for windows is made by floating molten glass on molten tin. Hydrogen removes oxygen. This stops the molten tin oxidizing or hazy residues forming on the glass.
Cryogenic Research	Hydrogen is used in cryogenic (low-temperature) research, including studies of superconductivity at low temperatures.
Minute Leak Detection	When mixed with nitrogen (sometimes called forming gas), can be used as a tracer gas.

Hydrogen can be produced through several methods. The most common methods are thermal, electrolytic, and photolytic processes:

- *Thermal Processes:* These processes allow the conversion of a variety of feedstocks.
  - *Steam Methane Reforming (SMR):* This is a high-temperature (700°C–1000°C) process in which steam reacts under 3–25 bar pressure and in the presence of a catalyst with a methane source (usually natural gas) to produce hydrogen, carbon monoxide, and a small amount of carbon dioxide. This is followed by a "water-gas shift reaction," where carbon monoxide and steam are reacted using a catalyst to

produce carbon dioxide and more hydrogen. The final process step is "pressure-swing adsorption," where carbon dioxide and other impurities are removed from the gas stream to result in pure hydrogen. The reactions require heat to be carried through. The SMR process can also be used to produce hydrogen from ethanol, propane, or even gasoline.

- *Partial Oxidation:* This process is similar to SMR, but takes place in the presence of a limited amount of oxygen. The process gives off heat, is faster than SMR, and requires a smaller reaction vessel.
- *Coal Gasification:* Coal is initially reacted with oxygen and steam under high pressure and temperature, forming a "synthesis gas" (mixture of carbon monoxide and hydrogen). The carbon monoxide in the synthesis gas is then reacted with steam via the water-gas shift reaction, creating additional hydrogen and carbon dioxide. Hydrogen is removed by a separation system. This process does produce a fair amount of carbon dioxide, which can be captured and sequestered.
- *Biomass Gasification:* This process is similar to coal gasification, but uses biomass as the feedstock. Biomass is converted into a gaseous mixture of mainly hydrogen, carbon monoxide, and carbon dioxide through the application of heat under pressure in the presence of steam and a controlled amount of oxygen.
- *Pyrolysis:* Gasification of biomass in the absence of oxygen.
- *Reforming of Renewable Liquid Fuels:* This is a process similar to reforming of natural gas, but involves the initial conversion of biomass resources to ethanol, bio-oils, or other liquid fuels. This provides the advantage of lower-cost transport of these liquid fuels to the point of use, where they can be reformed to produce hydrogen.
- *High-Temperature Water Splitting:* High-temperature (500°C–2000°C) heat is used to drive a series of chemical reactions, in a closed loop (chemicals are reused in each cycle), to produce hydrogen and oxygen. High-temperature heat can be supplied by nuclear reactors under development (up to about 1000°C) or solar concentrators (up to about 2000°C).

- *Electrolytic Processes:* Water can be separated into oxygen and hydrogen through a process called electrolysis, which takes place in an electrolyzer (functions like a fuel cell in reverse). Electricity cost is the primary price driver for hydrogen production through electrolysis. Increasing the electricity cost by approximately three times nearly doubles the production costs. In the low case, electricity accounts for over 60 percent of the total costs; in the high case, it is over 80 percent of the total production costs.

- *Photolytic Processes:* Photolytic processes use light as the agent for hydrogen production. Photobiological processes use the natural photosynthetic activity of bacteria and green algae to produce hydrogen. Photoelectrochemical processes use specialized semiconductors to separate water into hydrogen and oxygen.

Hydrogen not produced at the point of use needs to be delivered, in either gaseous or liquid form. Liquid hydrogen must be cryocompressed, which requires additional energy compared to gaseous hydrogen; however, the liquid form is denser, requiring less volume. Gaseous hydrogen also requires additional energy input for transport, as compression is necessary to efficiently transport the gas. Transporting hydrogen can be done via pipeline or truck.

Delivery of hydrogen is categorized as either small or large merchant. Small merchants transport hydrogen predominantly by liquid tanker (90 percent), with compressed gas tube trailer (7 percent) and compressed gas cylinders (3 percent) making-up the rest. Large hydrogen merchants tend to use hydrogen pipelines to transport their product. With the current advanced technologies available for shipping hydrogen, the cost of delivery from a central production facility to a fueling station 60 miles away is estimated to be $2–$3/gge.

For the smallest quantities (individual deliveries of 0.5–50 kg), hydrogen is often bought and sold as a compressed gas in cylinders, which allows it to be easily transported between the point of production and the point of use. These cylinders are most commonly made of steel and come in a variety of sizes and pressures. For larger quantities (50+ kg), tube trailers (with compressed gaseous hydrogen) and liquid tanker trucks (with cryogenic liquid hydrogen) are more likely to be used for delivery. Tube trailer quantities range from 100 kg for a "pup trailer" to 300 kg for large trailers. A common size in the middle is 165 kg. Liquid truck capacities range from about 2,300 to 4,300 kg shows common cylinder, tube trailer, and liquid trailer quantities.

For storing hydrogen, efforts are focused on finding a storage system to facilitate commercialization of fuel cell systems. This wide range of efforts is designed to develop large and small scale technologies which can provide hydrogen, dispensed and untaxed, for between $2 and $4/gge. Within this range, hydrogen could be competitive with gasoline. The main methods of hydrogen storage utilized, to improve the energy density of hydrogen, can be characterized as:

- *Gaseous Storage:* Hydrogen is stored in high-pressure (5,000–10,000 psi) compressed gas tanks.

- *Liquid Storage:* Energy density of hydrogen is further improved by storing in a liquid state. However, the energy requirement for liquefying hydrogen is high. Approximately 30 percent of the heating value of hydrogen is required for liquefaction.
- *Materials-based Storage:* Storage of hydrogen by binding to different materials through absorbing directly into the materials, adsorption (requiring highly porous materials to maximize the surface area available for hydrogen sorption to occur), and chemical reactions.

Once created, stored and delivered, the use of hydrogen happens via a fuel cell. Fuel cells are an emerging technology that can provide heat and electricity for buildings and electrical power for vehicles and electronic devices. Fuel cells work like batteries, but they do not run down or need recharging. They produce electricity and heat as long as fuel is supplied. A fuel cell consists of two electrodes, a negative electrode (or anode) and a positive electrode (or cathode), sandwiched around an electrolyte. Hydrogen is fed to the anode, and air is fed to the cathode. Activated by a catalyst, hydrogen atoms separate into protons and electrons, which take different paths to the cathode. The electrons go through an external circuit, creating a flow of electricity. The protons migrate through the electrolyte to the cathode, where they reunite with oxygen and the electrons to produce water and heat.

Although the basic operations of all fuel cells are the same, numerous varieties have been developed to take advantage of different electrolytes and serve different application needs:

- *Polymer Electrolyte Membrane Fuel Cells:* Polymer electrolyte membrane (PEM) fuel cells, also called proton exchange membrane fuel cells, use a polymer membrane as the electrolyte. These cells operate at relatively low temperatures and can quickly vary their output to meet shifting power demands. PEM fuel cells are the best candidates for powering automobiles. They can also be used for stationary power production.
- *Direct-Methanol Fuel Cells:* The direct-methanol fuel cell (DMFC) is similar to the PEM cell in that it uses a polymer membrane as an electrolyte. However, DMFCs use methanol directly on the anode, which eliminates the need for a fuel reformer. DMFCs are of interest for powering portable electronic devices, such as laptop computers and battery rechargers.
- *Alkaline Fuel Cells:* Alkaline fuel cells use an alkaline electrolyte such as potassium hydroxide or an alkaline membrane. Originally used by

NASA on space missions, alkaline fuel cells are now finding new applications, such as in portable power.

- *Phosphoric Acid Fuel Cells:* Phosphoric acid fuel cells use a phosphoric acid electrolyte held inside a porous matrix, and operate at about 200°C. They are typically used in modules of 400 kW or greater and are being used for stationary power production in hotels, hospitals, grocery stores, and office buildings, where waste heat can also be used. Phosphoric acid can also be immobilized in polymer membranes, and fuel cells using these membranes are of interest for a variety of stationary power applications.
- *Molten Carbonate Fuel Cells:* Molten carbonate fuel cells use a molten carbonate salt immobilized in a porous matrix as their electrolyte. They are already being used in a variety of medium-to-large-scale stationary applications, where their high efficiency produces net energy savings. Their high-temperature operation (approximately 600°C) enables them to internally reform fuels such as natural gas and biogas.
- *Solid Oxide Fuel Cells:* Solid oxide fuel cells use a thin layer of ceramic as a solid electrolyte. They are being developed for use in a variety of stationary power applications, as well as in auxiliary power devices for heavy-duty trucks. Operating at 700–1000°C with zirconia-based electrolytes, and as low as 500°C with ceria-based electrolytes, these fuel cells can internally reform natural gas and biogas, and can be combined with a gas turbine to produce electrical efficiencies as high as 75 percent.

The U.S. Department of Energy (DOE) has set a target of \$30/kW (by 2017) for transportation fuel cell systems [26]. Through successful R&D efforts, the cost of automotive fuel cells has been reduced by more than 35 percent since 2008 and by more than 80 percent since 2002: from \$275/kW in 2002 to \$47/kW in 2012, based on projections to high-volume (500,000 units per year) manufacturing [27]. This significant progress is shown in Figure 4.3.

Niche markets are considered the best early applications for hydrogen market integration, including specialty vehicles which have already proven fruitful. DOE is focusing its efforts on developing new technologies and materials to enable low-cost hydrogen generation from renewable resources. With hydrogen production, the goal is to develop distributed production technologies for fueling stations, avoiding the expenses associated with delivery and its required infrastructure. Longer-term efforts for small facilities focus on electrolysis and liquids reforming while research for larger capacity plants is focused on exploring hydrogen production from biomass, wind, and solar-powered processes.

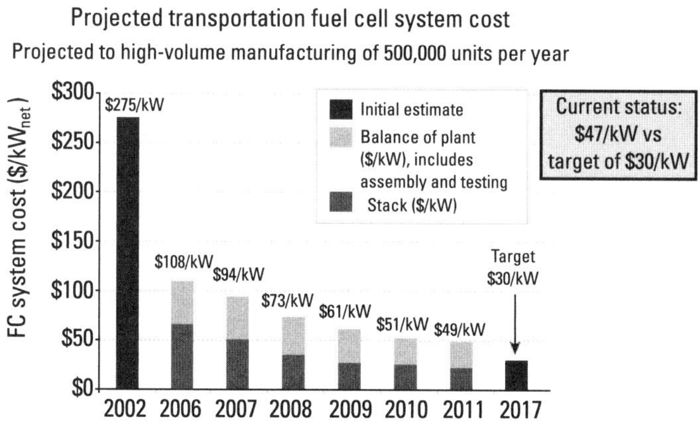

Figure 4.3 Advances in reducing transportation fuel cell system costs [26].

Early (current through 2014) emerging markets include hydrogen fuel cells for stationary power applications (both primary and backup), portable power, and lift trucks. In the mid-term, commercial markets for hydrogen fuel cells are expected to include residential combined heat and power (CHP) systems (2012 to 2017), auxiliary power units, and fleets and buses. Fuel cell technology has already been successfully integrated into specialty vehicle markets such as lift trucks and airport tugs. It is possible, with available technology, to generate hydrogen from natural gas at a cost which makes it competitive with gasoline. Assuming high-volume production onsite through natural gas reforming, a fueling station could provide hydrogen for approximately $3 per gallon gasoline equivalent (gge), including compression, storage and dispensing. High-volume production at the fueling station can provide hydrogen for approximately $4.90–$5.70/kg with existing technologies, including compression, storage and dispensing. Renewable hydrogen produced at a central plant from wind-powered water electrolysis costs $2.70–$3.50/kg when scaled to high volumes, excluding delivery, compression, storage and dispensing [28].

In the long term, 2015 to 2020 timeframe and beyond, hydrogen fuel cells are projected to expand their base in current markets and to enter the commercial light-duty vehicle (LDV) fleet. The Department of Energy's Controlled Hydrogen Fleet and Infrastructure Demonstration and Validation project has produced and/or dispensed 151,900 kg of hydrogen to fuel cell vehicles between the second quarter of 2005 and the third quarter of 2011 to further the realization of this goal [29].

Growing interest and investment among leading world economies underscores the global market potential for hydrogen and fuel cells. While fuel cells are becoming competitive in some markets, the range of these markets can be greatly expanded with further technical improvements and reductions in manufacturing costs. Additionally, certain institutional and economic barriers need to be overcome such as the need for codes and standards, the lack of public understanding of the technologies, the high initial costs, and the lack of a supply base that many new technologies face in their critical early stages. As with any new technology, hydrogen-related codes, standards, and safety measures, need to be further developed and harmonized at national and international levels.

Efforts also need to continue to keep increasing public knowledge of fuel cell and hydrogen systems, especially to overcome the general perception that hydrogen is unsafe and unreliable. The high expenses associated with hydrogen use will not be overcome without economies of scale and the corresponding domestic supplier base, product demand, and sales numbers. Economies of scale will not be reached with the uncertainty of political and financial support which currently exists. The successful commercial integration of hydrogen will require investments in advanced manufacturing technologies, generation capabilities, and efficient supply-chain networks to attain lower operating costs. In addition, simultaneous investment in delivery infrastructure and end-use technologies will be necessary to sustain and build this industrial sector.

4.2.8 Electricity

As technologies evolve and issues of oil dependence and climate change create more pressure, there is a gradual shift towards *electric drive* vehicles. Five main vehicle types may be categorized as electric drive: battery electric vehicles (BEV), plug-in hybrid electric vehicle (PHEV), fuel cell electric vehicle (FCEV), hybrid electric vehicle (HEV), and extended-range electric vehicle (EREV). While all these vehicle types operate on an electric platform and include batteries, there are some nuances to note.

BEVs and PHEVs utilize electricity from the grid by being plugged into an electric power source. A BEV, also may be called a "pure electric vehicle," is powered *only* by the electricity contained in its onboard battery. The main concept is the same as any other battery-operated device, the electricity stored in the battery takes the vehicle only so far, and needs to be recharged (by plugging-into the grid). On the other hand, a PHEV makes use of *both* an electric motor (that makes use of electricity stored in a battery and can be re-charged by plugging-in to the grid), and an ICE (that can utilize conventional or alternative fuels). EREVs are also essentially PHEVs, but utilize the gas engine exclusively for recharging batteries, instead of for directly driving the wheels (as in HEVs).

An HEV also uses an ICE *and* an electric motor, similar to a PHEV. The difference here, however, is that the battery is charged either by regenerative breaking, or by the ICE (*not* by plugging-in).

A FCEV is somewhat different than the other options, in that it uses a fuel, hydrogen. The chemical energy in hydrogen is converted to electrical energy through a fuel cell, and this electric energy drives the electric motor, to run the vehicle. Similar to conventional gasoline cars, once the hydrogen fuel runs out, the FCEV needs to be "refueled" with hydrogen at hydrogen refueling stations.

4.2.8.1 HEVs

The use of an electric motor in an HEV provides extra power, reducing the size of the ICE, and thus improving fuel economy. As driving is supported by the electric motor, tailpipe emissions are reduced. Consumer acceptance and interest in HEVs have also been sizeable, as it provides driving ranges similar to conventional vehicles, while also providing some level of environmental benefits, independence from oil, and buffer against rising oil prices. While the purchase price of HEVs are generally more than conventional vehicles, tpart of this difference may be offset through fuel savings and financial incentives for purchase. The *regenerative breaking* system in an HEV operates by the energy normally lost during braking being captured by using the electric motor as a generator and storing the captured energy in the battery. This regenerative breaking concept allows the HEV to be well-suited and beneficial for start-and-stop city traffic. While at highway cruising speeds, the HEV typically makes use of its conventional ICE. However, the energy stored in the battery also provides extra power for the vehicle's acceleration, as well as auxiliary power when the vehicle is idling.

HEVs can be either *mild* or *full* hybrids, and full hybrids can be designed in either a *series* or *parallel* configuration. *Mild hybrids* use a battery and electric motor to help power the vehicle (but cannot power vehicle using electricity alone) and can allow the engine to shut off when the vehicle stops at traffic lights or in stop-and-go traffic. This characteristic further improves fuel economy. While mild HEVs cost less than full HEVs, they provide lower fuel economy than full hybrids. *Full hybrids* can be powered by just electric power (for short distances and at low speeds), as they have more powerful electric motors and larger batteries. While they provide better fuel economy, the larger batteries required cause these vehicles to cost more than mild hybrids. *Parallel hybrids,* a category of full hybrids, are the most common HEV design. As the fuel tank supplies gasoline to the engine, a set of batteries supplies power to an electric motor, and both the electric motor and the gas engine can provide propulsion power. Series hybrids, another category of full hybrids, are actually a more common configuration for PHEVs. Here, the gasoline engine turns a generator, and the generator can either charge the batteries or power an electric motor that drives the transmission. Thus, the gasoline engine never directly powers the vehicle.

4.2.8.2 PHEVs

PHEVs use essentially the same technology as an HEV but use larger high-voltage batteries to power the electric motor and have a longer all-electric range. Greater use of the electric drive allows the vehicle to use less gasoline and thus become more efficient than conventional HEVs. Some PHEV retrofits have even achieved 100 mpg over 1,000 miles of travel [28]. But, unlike the batteries in the HEV, which are recharged by only the internal combustion engine, the batteries in PHEVs can also be recharged by plugging-in to an external power source. PHEVs are generally described by their all-electric range, which is the distance the vehicle can travel with minimal usage of the internal combustion engine for assistance. For example, a PHEV40 would mean the vehicle could travel 40 miles on all-electric mode. Current PHEV models are able to travel 10 miles to 40 miles on all-electric mode [31]. One advantage this plugging-in capability allows is that vehicles may be charged at night by grid power during off-peak hours, when electricity rates are typically lower than daytime peak rates. This night-time demand also can create a market for intermittent renewables like wind, which are in excess during night time and need to be curtailed.

4.2.8.3 BEVs

As BEVs *only* use electricity and no petroleum, they contribute more to reducing petroleum dependency and local air pollution due to tailpipe emissions. However, even though BEVs result in no tailpipe emissions, depending on what electricity source/mix is used to charge the vehicle, there most likely will be emissions from the production of electricity (unless a pure renewable option is used). Thus, on a life-cycle basis, there still are emissions involved. A more detailed analysis comparing BEVs to other vehicle options is shown in Section 4.3.1. Thus, the pollution is usually not necessarily completely avoided, but essentially shifted from many point sources to one or two central sources. The driving range of BEVs (on a full charge), are currently shorter than the 300 mile range offered by a tank of fuel in conventional gasoline vehicles (manufacturers typically target around a minimum of 100 miles [31]). However, research [32] compiled by the U.S. Department of Transportation's (DOT's) Federal Highway Administration has shown that a majority (63 percent) of trips by the average household are less than 10 miles. Longer tripsof 50 miles or more make-up only three percent of trips. Therefore, a BEV would fit the needs of many consumers in their daily commuting activities. Still, *range anxiety* (afraid of being stranded on the road due to depleting the car battery) is an issue for consumers. The range, as well as the established fueling infrastructure present for the conventional gasoline ICEV is something that consumers are accustomed to and take for granted. Thus, when considering alternative transportation technologies whether it is a BEV, PHEV, or FCEV consumers expect at least the level of comfort and reliability that has been provided to them over decades.

4.2.8.4 Charging of BEVs and PHEVs

Charging of BEVs and PHEVs is accomplished through plugging the vehicle in to an electric outlet, through the support of connectors, grounding, and other equipment necessary (collectively called "electric vehicle supply equipment;" EVSE). Charging of these vehicles takes much longer than the few minutes consumers are used to fueling their ICEVS. Depending on the vehicle type, how depleted the battery is, how much energy holding capacity the battery has, the type of battery, and the characteristics of the EVSE, charging time may vary between 30 minutes to more than 20 hours. Thus, charging will need to take place at locations where the vehicle is able to devote to being parked for some amount of time such as at home, workplaces, or other public parking spots [31]. Table 4.4 displays various EVSE characteristics and charging options.

4.3 Impacts

Section 4.3.1 compares the environmental, energy use, and cost impacts of various alternative fuels, while Section 4.3.2 explains an emerging issue related to ethanol called the "ethanol shuffle."

4.3.1 Well-to-Wheel Greenhouse Gas, Petroleum Energy Use, and Cost Impacts

A detailed well-to-wheels (WTW) analysis [33] was conducted by the National Renewable Energy Laboratory (NREL) to investigate various pathways to producing, delivering, and distributing hydrogen for use in fuel cell vehicles. Emissions, energy use, and costs of these pathways were evaluated and compared

Table 4.4
Electric Vehicle Supply Equipment Characteristics and Charging Options [31]

	Current Type	Amperage (amps)	Voltage (V)	Kilowatts (kW)	Charging Time	Primary Use
Level 1	Alternating Current (AC)	12–16	120	1.3–1.9	2–5 miles of range per hour of charging	Residential Charging
Level 2	AC	Up to 80	240	Up to 19.2	10–20 miles of range per hour of charging	Residential and Public Charging
Level 3 *	AC	To be determined	To be determined	To be determined	60–80 miles of range in less than 30 minutes	Public Charging
DC Fast Charging	Direct Current (DC)	Up to 200	208–600	50–150	60–80 miles of range in less than 30 minutes	Public Charging

*In development.

against gasoline, diesel, and ethanol-based vehicles. The seven pathways for hydrogen production include:

- Two distributed (onsite) production options:
  - Hydrogen produced from natural gas ("distributed natural gas"), and distributed as 350 bar compressed gas.
  - Hydrogen produced from grid electricity ("distributed electrolysis"), and distributed as 350 bar compressed gas.
- Five central production options:
  - Hydrogen produced from biomass, delivered as liquid hydrogen in trucks, and distributed as 350 bar compressed gas.
  - Hydrogen produced from biomass, delivered as gaseous hydrogen in pipelines, and distributed as 350 bar compressed gas.
  - Hydrogen produced from natural gas, delivered as gaseous hydrogen in pipelines, and distributed as 350 bar compressed gas.
  - Hydrogen produced from wind electricity, delivered as gaseous hydrogen in pipelines, and distributed as 350 bar compressed gas.
  - Hydrogen produced from coal (with carbon sequestered), delivered as gaseous hydrogen in pipelines, and distributed as 350 bar compressed gas.

The WTW GHG and levelized fuel cost per mile for the seven hydrogen pathways were compared with the WTW GHG emissions and market price per mile for gasoline, diesel, and E85. The two extremes were found to be the distributed electrolysis and central wind electricity (with pipeline delivery) options. The distributed electrolysis case has very high GHG emissions compared to the other options, due to the nature of the grid mix assumed (mostly fossil-based U.S. average grid electric mix). The wind electricity case is amongst the lowest in GHG emissions, but has the highest cost, due to the status of the technology. For the rest of the hydrogen pathways, the per mile costs are within the ranges of the nonhydrogen options, and the hydrogen options provide about equal (as compared to gasoline HEVs) or lower GHG emissions.

As part of the analysis, for the same range of options, NREL analysts also looked at the effects of various carbon taxation levels, if they were to be implemented. The tax calculation took into account the GHG emissions from the whole life cycle of the pathways, so both upstream and production process emissions were considered. It was found that at a \$100/ton CO_2-equivalent tax, the per-mile levelized cost of hydrogen would be increased by 81 percent for the distributed electrolysis pathway (due to the coal-based U.S. electric grid amounting to a lot of carbon emissions), 40 percent for the central natural gas pathway, and 43 percent for the distributed natural gas case, while increasing the cost of all other pathways by less than 20 percent. When compared with the non-hydrogen pathways, the effects of the two different levels of the carbon tax vary, in some cases putting some hydrogen pathways at a disadvantage, while in others improving their stance.

The U.S. DOE has collaborated with national labs, researchers, and analysts, to conduct a comprehensive assessment of a portfolio of vehicle technologies used in conjunction with a variety of fuel pathways to determine the WTW effects on GHG emissions (grams of CO_2-equivalent per mile) and petroleum use (Btus per mile) of these options. Some of the main assumptions of the study were that: analyses are for a projected state of technologies in 2035 to 2045; ultra-low carbon renewable electricity includes wind, solar, etc.; and analyses do not include the life-cycle effects of vehicle manufacturing and infrastructure construction/decommissioning. More detailed explanations of all the assumptions may be found in DOE's Hydrogen and Fuel Cells Program Record [34].

The analysis revealed that hydrogen fuel cell vehicles have the potential to provide 55 percent reduction in GHG emissions compared with today's gasoline vehicles. When compared to future hybrid electric vehicles, fuel cell vehicles utilizing renewable-based hydrogen could reduce CO_2 emissions by up to 84 percent. In addition, hydrogen-based fuel cell vehicles may reduce oil consumption by up to 80–95 percent, as compared with today's gasoline internal combustion vehicles, advanced hybrid electric vehicles (running on gasoline or ethanol), and advanced plug-in hybrid electric vehicles.

4.3.2 The "Ethanol Shuffle" Concept

A phenomenon in ethanol trade called the "ethanol shuffle" emerged at the end of 2011, related to the U.S. Environmental Protection Agency (EPA) and California Air Resources Board (CARB) calculations of carbon emissions for corn-based ethanol and Brazilian sugar ethanol. The carbon footprint of Brazilian based sugar ethanol was deemed far superior to corn-based ethanol. This resulted in growing incentives for imports of ethanol from Brazil, while the struggling Brazilian ethanol industry could not meet its own domestic demand. Brazilian ethanol producers were finding it more valuable to export their product to the United States and to import growing volumes of U.S. ethanol. However, this export-import activity also caused additional carbon emissions due to all the ocean transport involved. So, California would be importing sugarcane ethanol from Brazil rather than corn ethanol from Nebraska or Kansas; and in turn, corn ethanol from the Midwest would travel to Houston or Galveston via rail, then would be shipped to Brazil via tanker to "backfill" the volumes they sent to the United States. All of this was being done in the name of "reducing GHG emissions," but ended-up causing negative environmental effects.

Transportation-related GHG emissions more than double in the scenario where California imports Brazilian cane ethanol and Brazil "backfills" those volumes with U.S. corn ethanol imports. Miles traveled in this scenario are more than eight times the miles traveled in a scenario where California ethanol demand is met with corn ethanol from the Midwest. In addition, there are economic effects. California gasoline blended with imported Brazilian ethanol has been 16 cents per gallon more expensive than gasoline blended with U.S. ethanol.

4.4 Market Status

The current market status and projected market trends of various alternative fuels are described in the following sections.

4.4.1 Liquefied Petroleum Gas

Globally, LPG is the third most used transportation fuel, with great popularity in countries like Japan, Canada, Italy and Austria [5]. In the United States, in 2009, there were 147,030 vehicles running on LPG in use. While use of LPG in vehicles increased steadily from 172,806 in 1995 to 190,369 in 2004, the trend since then has been a decrease in use [35]. As of May 2012, approximately 2,661 LPG refueling sites were operational in the United States, with the majority located in Texas (480 stations), followed by California (228 stations), and Indiana (183 stations) [1].

4.4.2 Natural Gas

In the United States, demand for natural gas vehicles (NGVs) is dominated by fleets. As of May 2012, stations dispensing CNG totaled 1,014, with the majority in California (228 stations), followed by New York (107 stations) [1]. It is projected that as fueling for fleets increases, consumer confidence will also increase, and up to 1,972 stations and sales of 32,619 NGVs could be seen in the U.S. by 2016.

Globally, the sale of NGVs is expected to continue to increase, reaching over 3 million NGVs sold globally by 2016, dominated by the markets in Asia and the Pacific. Overall, the NGV vehicle market globally is expected to experience a compound annual growth rate of 9.1 percent between 2010 and 2016, which is projected to translate to 6.7 compound annual growth rate in the global use of natural gas as a transportation fuel, with total usages reaching 19,123 million cubic feet by 2016 [36].

In Latin America, Brazil, and Argentina dominate the natural gas vehicle market, with a combined total of NGVs that amount to 28 percent of the NGV vehicles worldwide. In Western Europe, Italy leads the NGV market but the European market varies according to country, with weaker markets in countries like Spain and the U.K. Most of the countries in Europe prefer to focus more on other alternative fuels, as a majority of natural gas in Europe has to be imported. In Asia and the Pacific region, India has become a strong player with a push for NGVs in transit fleets resulting in India having the largest number of CNG bus fleets globally. CNG is also gaining traction in the passenger vehicle market in India, due to the lower cost of the fuel and availability of CNG stations. China has typically focused more on the use of natural gas for industrial applications, but has recently been moving towards the adoption of CNG buses.

4.4.3 Ethanol

In the United States, the ethanol industry has grown. In 2011, a total number of 209 biorefineries in 29 states generated 13.9 billion gallons of ethanol, a far cry from the 1.63 gallons produced just a bit over a decade ago in 2000. The economic crisis of 2008 caused several biorefineries to cease operations, but by 2011, they had entered the playing field once again. Additionally, most of the operational biorefineries are updating the technologies they employ, resulting in improved process efficiency and quality of by-product feed produced. The corn oil created during the ethanol production process serves as a valuable, additional revenue-generating by-product and 40 percent of the ethanol biorefineries in the United States are selling this oil to various markets (feed, biodiesel, other chemical) [3].

In terms of ethanol trade, the United States exported over one billion gallons ethanol (representing an increase of 600 million gallons over 2010) in 2011, mainly to Brazil, Europe, and Canada with some amounts also exported to the Middle East and Asia. The United States also exported nearly 8 million metric tons of distillers' grains in 2011, to markets in North America, Asia and Europe [3].

The global ethanol supply in 2011 was dominated by the United States (56.7 percent), followed by Brazil (29.8 percent). The remaining 13.5 percent of global supply is distributed mainly between France (2.5 percent), Canada (2.3 percent), China (2.2 percent), Germany (1.9 percent), India (1 percent), and other countries (3.7 percent). Global ethanol demand followed a similar trend, with again the majority demanded by the United States (54.7 percent), followed by Brazil (26.9 percent). The remaining demand centers were France (3.1 percent), Canada (2.9 percent), China (2.7 percent), Germany (3.8 percent), India (1 percent), and other countries (4.8 percent) [6].

Over the past 13 years, U.S. ethanol production and consumption has remained almost equal [1]. Globally, ethanol production is dominated by markets in the United States and Brazil [1].

E85 stations in the United States are concentrated in the middle of the country (as corn, the major ethanol feedstock, is grown there). Minnesota has the most number of stations, followed by Illinois [1].

In the United States, in January 2011, EPA approved the use of ethanol blend E15 for use in light duty cars, trucks, and sport utility vehicles model year 2001 and newer and all flex-fuel vehicles. Use of E15 in the United States could create a nationwide annual ethanol demand of close to 20 billion gallons, which would in turn help with meeting the requirements of federal standards like the Renewable Fuel Standard (RFS), while creating the market demand that will encourage investment in cellulosic ethanol technologies [2].

A major incentive for the U.S. ethanol market has been the Volumetric Ethanol Excise Tax Credit (VEETC), which was created by the American Jobs Creation Act of 2004. This incentive provided blenders and marketers of fuel with a federal tax credit of 45 cents on each gallon of ethanol blended with their gasoline and did not distinguish between feedstocks. Thus, all types of ethanol, including ethanol produced from new cellulosic feedstocks received the benefits of the credit. In December 2010, the VEETC was extended for an additional year as part of a tax rate compromise reached between President Obama and Congress. However, the credit was allowed to expire by December 31, 2011. While in force, the VEETC had reduced oil imports in the United States by 485 million barrels a year and had also resulted in the creation of 401,600 domestic jobs. Thus the extension of current incentives such as the Production Tax Credit for cellulosic ethanol and the Accelerated Depreciation for Cellulosic Biorefineries is of great importance.

4.4.4 Biodiesel

The production, consumption, and export activities of biodiesel in the United States started gaining significance around 2004, and especially after 2006. The year 2007 is when export activities emerged, increasing in 2008, but have been decreasing since then. Domestic biodiesel production and consumption levels have followed a similar trend, until 2011, when both demonstrated significant increases of well over 100 percent. Most of the production was consumed domestically, and exports decreased by approximately 30 percent, in comparison to 2010 [1].

Biodiesel (B20 blend or higher) stations in the United States are distributed throughout the country. North Carolina has the most stations, followed by California, while West Virginia currently has no biodiesel supplying stations [1].

4.4.5 Algae-Based Biodiesel

A market analysis [9] conducted by Pike Research in 2010 projected that the global algae-based biofuels market would grow at a CAGR of 72 percent, reaching 61 million gallons of production per year by 2020, with a market value of $1.3 billion. Pike has observed that pilot projects have already started in the United States, allowing the country to become an early player in the global market in ramping-up production levels. The European Union (EU) is expected to initially be constrained to more emphasis on university research, and later by inadequate water, nutrient, and land resources. Latin America and the Asia Pacific regions are projected to grow significantly in the long-term, due to abundant land and water resources, as well as access to cheap labor. While the Middle East region and Africa have vast amounts of desert area suitable for algae production, the growth of the algae market in these areas is expected to be limited by preference of petroleum, political instability, infrastructure issues, and inadequate access to capital. In terms of the projected growth of algae-based biofuels in different regions of the world through 2020, production has nearly doubled in the last couple years of the decade. The North American market has displayed significant growth, and the Asia Pacific region demonstrated rapid growth in the last two years of the decade.

4.4.6 Hydrogen

As the hydrogen infrastructure for transportation is still emerging, the network of hydrogen stations in the United States is not as developed and widely spread as some other alternative fuels. Most development is concentrated in California, while some other states have 10 stations or less, and quite a few do not have any stations at all [1].

Pike Research [37] has forecast that worldwide demand for hydrogen will be more than 400 million kg annually, representing an 88 percent CAGR. Initially, the global demand for hydrogen is expected to be lower (reaching only 50 million of the 400 million kg) between 2010 and 2014, dominated by forklifts and uninterruptible power sources (UPS). It is then forecast to increase significantly, prompted by the expected introduction of commercial hydrogen fuel cell LDVs around 2015.

FCEV commercialization is expected to be in the hundreds of units in 2012 and in the thousands of units by 2015, led mainly by Germany, Japan, and Korea. Japan has unveiled a plan to sell two million FCEVs by 2025 and to install 1,000 hydrogen fueling stations, beginning with 50 to 100 stations in four metropolitan areas, which would then be linked by stations on intercity highways. China's latest five-year energy plan also includes hydrogen and fuel cells. GM has introduced its "production intent" FCEV and restated its plan to introduce a commercial vehicle by 2015.

4.4.7 Electricity

The National Research Council (NRC) of the National Academies conducted an analysis in 2010 [38] to investigate the status of PHEVs, consider factors that would affect their market penetration, and to determine a maximum practical penetration rate for PHEVs in the United States, along with estimates of costs and impacts on petroleum consumption and CO_2 emissions. Some of the key conclusions that this study reached were:

- While battery costs have displayed reductions as the technology develops, they remain high and the potential for significant reductions seems limited, with an expectation that costs will decline by approximately 35 percent by 2020 but at a slower rate afterwards.
- The incremental cost of a PHEV is expected to range from $5,500 to $6,300 (plus, some homes may require electrical system upgrades costing over $1,000), compared to an incremental cost of an HEV of $3,000.
- By 2030, 40 million PHEVs might be on the road. However, factors like high battery costs, limited charging points, consumer resistance, and modest gasoline savings are expected to result in this number being lower.
- PHEVs are not expected to have as much of an impact on oil consumption before 2030, based on market penetration, but more sizable reductions in oil use may be achieved by 2050.
- While PHEV10s are expected to emit less CO_2 than conventional vehicles, the savings they provide in comparison to HEVs will be relatively small considered on a life-cycle basis, when including emissions at electric generating stations supplying the power.
- Generation and transmission of power for millions of PHEVs over the next couple of decades should be adequate if charging is done at night.

Pike Research conducted a study [39] of the global market for electric vehicles (to include light-duty hybrid, plug-in hybrid, and battery electric vehicles) for the period 2011 through 2017. Pike's assessment was that the most rapid growth in the number of BEV and PHEV models available will likely be in the Asia Pacific region, followed by Europe and North America. HEVs will tend to be favored more in North America. An important factor in the development of these advanced vehicle market is the availability of supporting incentives. According to Pike, both China and the Asia Pacific region as a whole are already providing strong support through incentives. In terms of the overall global market growth for electric-bases vehicles, a CAGR of 19.5 percent (between 2011 and 2017) is expected, a nontrivial amount considering that the global vehicle market in general (all types of vehicles) is expected to grow at a CAGR of only 3.7 percent during the same timeframe. Specifically the PHEV market is anticipated to reach 1.4 percent of the global LDV market, with the Asia Pacific region in the lead (due mostly to growth in China and Japan). However, Pike analysts have concluded that the global PHEV market might not necessarily meet all of the deployment and commercialization targets set by various governments, based on the pace of current deployments. In terms of the global mix of HEVs, PHEVs, BEVs, and the trends in sales from 2001 to 2017, while HEV sales are expected to tend to remain fairly constant (showing only a slight increase) from 2014 to 2017, PHEVs, and especially BEVs are expected to gain a bigger share of the global market.

In 2010, Pike Research conducted a market study [40] focused on the potential and outlook of electric vehicles (PHEVs and BEVs) for the Asia Pacific Region. Pike forecasted that the Asia Pacific region—will be the main market for electric vehicles due to demand in China, Japan, and Korea. Between 2010 and 2015, sales of electric vehicles in the Asia Pacific region were forecast to have a 28 percent CAGR and exceed 1.2 million by 2015. Within the region, China is expected to experience the fastest growth, with a CAGR of 76 percent. Pike also projected that charging stations in this region would reach more than 860,000 units between 2010 and 2015, demonstrating a CAGR of 91 percent and generating revenue of over $865 million by 2015.

The two main drivers behind interest in electric vehicles in the Asia Pacific region are oil prices and climate change concerns. Major countries in the region have been developing deployment roadmaps and setting carbon dioxide emissions standards, while increasing various incentives and subsidies to encourage consumers. China is becoming an important player in the electric vehicle market, causing anxiety with competitors in Japan and Korea, while creating the capability to impact global markets. China already has approximately 55 manufacturers developing electric vehicles and the country is expected to develop into one of the largest markets for electric vehicles. In addition, the chemicals and electronics industry in China is gearing-up for this demand, by growing their rechargeable battery business. In Japan, the focus is on advancing the market for vehicle charging infrastructure. Japan has set targets to deploy 32,000 EVs and PHEVs, as well as 5,000 charging points by 2013 [40].

Electric vehicle charging stations throughout the United States vary between states and regions. California is currently the leader, with over 500 stations. Other strong deployment areas are Texas, Florida, and Washington [1].

4.5 Challenges and Potential Solutions

The challenges that biofuels, hydrogen, and electricity (as a fuel) are depicted in the following sections.

4.5.1 Biofuels

In general, the cost of producing biofuels such as cellulosic ethanol and biodiesel is currently higher than the cost of conventional petroleum-derived gasoline, remaining a significant barrier to increased market penetration. While the price of gasoline is determined primarily by the market price of oil, the largest cost in the production of biofuels is the cost of the feedstock. Once produced, biofuels cannot be delivered via the existing transportation infrastructure and therefore must be transported by more expensive options. High feedstock costs and a lack of infrastructure are significant barriers to an established biofuels market, and to become competitive costs will have to drop along the entire technology supply chain [37].

The International Energy Agency reports an average range of production costs for biofuels in 2010. These values, along with their projected future values, are given below in Table 4.5. As shown, production costs for ethanol (produced via fermentation) and biodiesel (via transesterification or hydroprocessing) currently cannot compete with petroleum gasoline production costs. Over time, these biofuel processes are expected to become much more competitive.

4.5.2 Hydrogen

Hydrogen fuel cell vehicles face the challenges of the chicken and egg—the dilemma of which comes first, the hydrogen infrastructure, or fuel cell vehicles?

Table 4.5
Global Average Production Cost Forecasts for Petroleum Gasoline and Various Biofuels [42]

Production Costs (USD/ liter gasoline equivalent)	2010	2015	2030	2050
Petroleum Gasoline	0.54	0.72	0.82	0.83
Ethanol—conventional	0.70–0.75	0.70 – 0.80	0.65 – 0.85	0.65 – 0.85
Ethanol—cane	0.60–0.70	0.60 – 0.70	0.60 – 0.70	0.60 – 0.70
Ethanol—cellulosic	1.05–1.15	0.90 – 1.05	0.80 – 0.95	0.80 – 0.90
Biodiesel—conventional	0.95–1.05	0.95 – 1.10	0.95 – 1.15	0.95 – 1.15
Biodiesel—advanced (BtL)	1.05–1.15	0.90 – 1.05	0.80 – 1.00	0.75 – 0.90

Note: Costs reflect global averages. Ranges result from the strength of correlation between oil price and feedstock costs and capital costs. Lower production costs result from a weak correlation, whereas higher costs result from a stronger impact of oil prices on feedstocks (20% impact) and capital costs.

The realization of a hydrogen-based transportation system is hard to come by if supply and demand is not well matched and optimized. If fuel cell vehicles hit the roads without enough hydrogen fueling stations to meet demand, or conversely, if an extensive hydrogen infrastructure is developed but fuel cell vehicles are not rolled-out in time, serious setbacks to development will occur. There could be consumer anxiety (not enough fuel to meet their demand), or valuable resources could be wasted on underutilized stations (not enough fuel cell vehicles to provide demand for this supply). Thus, the development and roll-out of hydrogen infrastructure and fuel cell vehicles requires a concerted and well-planned effort among the parties involved. Major automakers and energy companies around the world are converging on common goals and signing agreements in an effort to develop markets for hydrogen-based transportation. Seven automakers (Daimler, Ford, GM/Opel, Honda, Hyundai/KIA, Renault/Nissan, and Toyota) signed a letter of understanding with energy companies and government agencies in 2009, asserting that a significant number of hydrogen FCEVs could be commercialized starting in 2015, with the help of supporting infrastructure to be developed [43]. Certain countries and regions have responded to this assertion by developing initiatives and commitments to create these markets. The "H2 Mobility" initiative was formed in 2009 in Germany, committing to developing a nation-wide hydrogen fueling network between 2012 and 2015, in support of 100,000 FCEVs per year. In 2009/2010, Japan developed a plan for 1,000 hydrogen fueling stations (through the cooperation of 13 domestic oil and gas companies) to be built and two million FCEVs to be deployed by 2025 [44].

4.5.3 Electricity

In 2010, the Electric Power Research Institute (EPRI) conducted a survey to determine the number of consumers, in an effort to try to characterize interest in electric-based vehicles [42]. This survey identified some key adoption barriers that need to be overcome for BEVs and PHEVs to penetrate the market. Consumers had anxieties related to charging availability, costs, vehicle range, battery maintenance/disposal, and safety. Most consumers felt that the full charging infrastructure support would not be in place, and that long charging times would cause distress during times of emergency (when the vehicle would be needed but not available due to lack of charge). Emergency situations were regarded as a concern when considering vehicle range too. Consumers were found to be open to charging their vehicles at home, especially during the nighttime hours, as it would fit their lifestyles and allow them to make use of lower nighttime electricity rates. However, consumers who do not have access to an electric outlet in a garage (those who are in apartments or who park on the street) viewed the situation as a barrier and felt that they would be more hesitant in purchasing an electric vehicle. One of the misconceptions amongst those surveyed was that they associated electric vehicles with small, under-performing vehicles that only the wealthy or "tree-huggers" would buy.

4.6 Emerging Opportunities

Biobutanol and biogas are emerging alternative fuels that are entering the market, as explained in the following sections.

4.6.1 Biobutanol

Biobutanol is a crop-derived alcohol which can be used similarly to ethanol in internal combustion engines. The benefit of biobutanol is its higher compatibility with the gasoline infrastructure; it can use the existing gasoline transport infrastructure while ethanol cannot. Biobutanol can be used in ICEs independently, as a blend with petroleum gasoline, or with ethanol. The Environmental Protection Agency already allows biobutanol to be blended into gasoline as an oxygenate at levels up to 11.5 percent by volume. Its drawback is its lower energy content: biobutanol has an energy content 10–20 percent lower than that of petroleum gasoline. Current efforts in the development and commercialization of biobutanol include a joint effort by DuPont and BP, and the work of a small company named ButylFuel, LLC. With the help of a DOE Small Business Technology Transfer grant, ButyFuelis developing a process whereproduction of the biofuel could be competitive with present-day petrochemical production processes.

4.6.2 Biogas

Biogas (also called biomethane or renewable natural gas) is a resource that is emerging as an alternative fuel, and as a resource for clean electricity production. Biogas may be produced, through the decomposition of organic matter in the absence of oxygen (anaerobic digestion), from the organic matter present in a variety of sources like sewage, animal by-products, and municipal solid waste in landfills. The methane content in biogas is around 50 percent to 80 percent, compared to the approximately 70 percent methane content of natural gas [1]. After being processed to certain purity levels required by vehicles, biogas may be substituted for natural gas, as an alternative fuel for natural gas vehicles, via distribution in existing natural gas pipelines.

While the benefits of the use of biogas are similar to the benefits of natural gas as an alternative fuel, the use of biogas does provide some additional benefits. Biogas can be considered a renewable resource, as opposed to natural gas. At the same time, the use of biogas prevents the release of methane (which has climate effects 21 times stronger than CO_2) into the atmosphere (the amount that would be released through activities and processes at landfills, wastewater treatment plants and animal waste management facilities). Additionally, the anaerobic digestion process reduces landfill waste and odors, produces nutrient-rich liquid fertilizer, and requires less land than aerobic composting.

Globally, the use of biogas as an alternative fuel is gaining importance. Sweden utilizes biogas in more than half of its 11,500 natural gas vehicles, while Germany and Austria are targeting 20 percent biogas as a vehicle fuel [1].

References

[1] Website: U.S. Department of Energy, Office of Energy Efficiency and Renewable Energy, Alternative Fuels Data Center, Available at: http://www.afdc.energy.gov/.

[2] Renewable Fuels Association, *2012 Pocket Guide to Ethanol,* February 2012.

[3] Renewable Fuels Association, *Accelerating Industry Innovation: 2012 Ethanol Industry Outlook*, February 2012.

[4] Website: NGV Journal, Worldwide NGV Statistics, Available at: http://www.ngvjournal.dreamhosters.com/en/statistics/item/911-worldwide-ngv-statistics (accessed May 2012).

[5] Selim, M. Y. E-S., "Liquefied Petroleum Gas." In *Alternative Fuels for Transportation,* pp. 203–226, A.S. Ramadhas (ed.), Boca Raton, FL: CRC Press, 2011.

[6] Rose, H., *Exporting Ethanol: Market Opportunities and Global Supply*, Presentation given at 2011 Southeast Bioenergy Conference, Tifton, GA, August 9, 2011.

[7] Lawrence, M., and C. Wheelock, *Biofuels Markets and Technologies, Research Report, 4Q 2011*, Pike Research, 2011.

[8] Organisation for Economic Cooperation and Development/International Energy Agency, *EV City Casebook: A Look at the Global Electric Vehicle Movement*, 2012.

[9] Lawrence, M., and C. Wheelock, Algae-*Based Biofuels, Research Report, 4Q 2010*, Pike Research, 2010.

[10] U.S. Department of Energy, Vehicle Technologies Program, *Propane Basics, Fact Sheet*, March 2010.

[11] U.S. Department of Energy, Vehicle Technologies Program, *Natural Gas Basics, Fact Sheet*, April 2010.

[12] Rao, G. L. N., and A. S. Ramadhas, "Compressed Natural Gas," *Alternative Fuels for Transportation*, pp. 227-242, A.S. Ramadhas (ed.), Boca Raton, FL: CRC Press, 2011.

[13] Canakci, M., and O. Ilgen, "Methanol," *Alternative Fuels for Transportation*, pp. 81–128, A.S. Ramadhas (ed.), Boca Raton, FL: CRC Press, 2011.

[14] U.S. Department of Energy, *Assessment of Costs and Benefits of Flexible and Alternative Fuel Use in the U.S. Transportation Sector, Technical Report No. 7: Environmental, Health, and Safety Concerns*, October 1991.

[15] U.S. Environmental Protection Agency, *Chemical Summary for Methanol, Report No. EPA 749-F-94-013a*, 1994.

[16] Hansen, A. C., C. E. Goering, and A. S. Ramadhas, "Ethanol," *Alternative Fuels for Transportation*, pp. 129–166, A. S. Ramadhas (ed.), Boca Raton, FL: CRC Press, 2011.

[17] Website: Greater New Haven Clean Cities, Inc., *Ethanol Vehicles*, Available at: http://nhcleancities.org/vEthanol.html

[18] International Energy Agency, *Biofuels for Transport: An International Perspective*, Paris, France: International Energy Agency, 2004.

[19] National Renewable Energy Laboratory, *Biodiesel Handling and Use Guide (Fourth Edition), NREL/TP-540-43672*, Golden, CO: National Renewable Energy Laboratory, December 2009.

[20] Ramadhas, A. S., S. Jayaraj, and C. Muraleedharan, "Biodiesel," *Alternative Fuels for Transportation*, pp. 41-80, A.S. Ramadhas (ed.), Boca Raton, FL: CRC Press, 2011.

[21] Hussain, K., et al., "Economically Effective Potential of Algae for Biofuel Production," *World Applied Sciences Journal*, Vol. 9, No. 11, 2010, pp. 1313–1323.

[22] U.S. Department of Energy, Biomass Program, National Algae Biofuels Technology Roadmap, May 2010.

[23] Darzins, A., P. Pienkos, and L. Edye, *Current Status and Potential for Algal Biofuels Production, Report to IEA Bioenergy Task 39, Report T39-T2*, August 6, 2010.

[24] Sikes, K., M. V. Walwijk, and R. McGill, "Algae as a Feedstockfor Biofuels: An Assessment of the State of Technology and Opportunities, Report from the IEA Advanced Motor Fuels Implementing Agreement," May 2011.

[25] Website: Praxair, Hydrogen Industrial Applications Information from Praxair, Industrial Gas Supplier, Available at: http://www.praxair.com.

[26] U.S. Department of Energy, *Fuel Cell Technologies Program, Multi-Year Research, Development, and Demonstration Plan: Fuel Cells*, March 2012.

[27] U.S. Department of Energy, *Hydrogen and Fuel Cells Program Record 12020: Fuel Cell System Cost-2012*, September 14, 2012.

[28] U.S. Department of Energy, Hydrogen and Fuel Cells Program Plan, September 2011.

[29] National Renewable Energy Laboratory, Hydrogen Fuel Cell Electric Vehicle Learning Demonstration, CDP#26: Cumulative Hydrogen Produced or Dispensed, December 9, 2011.

[30] Randolph, J., and G. M. Masters, *Energy for Sustainability: Technology, Planning, Policy*, Washington, D.C.: Island Press, 2008.

[31] U.S. Department of Energy, *Vehicle Technologies Program, Hybrid and Plug-In Electric Vehicles, Fact Sheet*, October 2011.

[32] U.S. Department of Transportation, Federal Highway Administration, *Our Nation's Hyways 2011, FHWA-PL-11-028*, 2011.

[33] Ruth, M., M. Laffen, and T.A. Timbario, "Hydrogen Pathways: Cost, Well-to-Wheels Energy Use, and Emissions for the Current Technology Status of Seven Hydrogen

Production, Delivery, and Distribution Scenarios, NREL/TP-6A1-46612," Golden, CO: National Renewable Energy Laboratory, September 2009.

[34] U.S. Department of Energy, "Hydrogen and Fuel Cells Program Record 10001: Well-to-Wheels Greenhouse Gas Emissions and Petroleum Use for Mid-Size Light-Duty Vehicles," October 25, 2010.

[35] Davis, S. C., S. W. Diegel, and R. G. Boundy, *Transportation Energy Data Book (30th Edition)*, ORNL-6986, Oak Ridge, TN: Oak Ridge National Laboratory, June 2011.

[36] Hurst, D., and C. Wheelock, *Natural Gas Vehicles, Research Report*, 1Q 2011, Pike Research, 2011.

[37] Jerram, L., and K-A. Adamson, *Hydrogen Infrastructure, Research Report*, 3Q 2011, Pike Research, 2011.

[38] National Research Council of the National Academies, *Transitions to Alternative Transportation Technologies—Plug-in Hybrid Electric Vehicles*, Washington, D.C.: The National Academies Press, 2010.

[39] Hurst, D., and J. Gartner, *Electric Vehicle Market Forecasts, Research Report*, 3Q 2011, Pike Research, 2011.

[40] Bae, H., J. Gartner, and D. Hurst, *Electric Vehicles in Asia Pacific, Research Report*, 4Q 2010, Pike Research, 2010.

[41] U.S. Department of Energy, Office of Energy Efficiency and Renewable Energy, Biomass Multi-Year Program Plan, April 2011.

[42] International Energy Agency, *Biofuels for Transport Roadmap*, 2011.

[43] International Partnership for Hydrogen and Fuel Cells in the Economy, "2010 Hydrogen and Fuel Cell Global Commercialization and Development Update, Fact Sheet," November 2010.

[44] Electric Power Research Institute and Southern California Edison, "Characterizing Consumers' Interest in and Infrastructure Expectations for Electric Vehicles: Research Design and Survey Results," May 2010.

5

Energy Storage

5.1 Basics

The storage of electric energy involves the use of kinetic, potential, or chemical energy forms to initially store the energy, later converting it to electricity. Energy storage serves various important functions: balancing fluctuations in electricity supply and demand; providing for peak electricity needs by using electricity stored during times of lower demand; and avoiding or delaying the need for the expansion of the electric grid [1]. The challenge of providing reliable services with variations in energy demand simultaneously has led to the development of various energy storage technologies. To ensure grid reliability, additional conventional generating units such as GHG emitting combustion turbines will need to be built, or nongeneration resources such as energy storage need to be integrated into the existing grid infrastructure. Energy storage could serve as a more effective way of meeting the increasing demand for ancillary services, with lower costs and less environmental impacts [2].

Various technologies for energy storage have been supporting electric energy systems for decades, such as pumped-hydro, compressed air energy storage (CAES), and lead-acid batteries. Newer energy storage technologies are now entering the market, such as lithium-ion batteries, flow batteries, flywheels, sodium-sulfur batteries, and hydrogen-based energy storage systems. These emerging technologies offer improved operational flexibility, better charge/discharge cycle life, and in some cases longer duration or fast response capabilities [3].

The performance and costs of energy storage systems are closely linked with the service they are to provide. For example, while back-up power systems require immediate (within seconds) response and moderate power output, and can usually handle higher costs (due to the high reliability provided), grid voltage and frequency regulation applications require a response time from seconds to minutes and higher power output. For daily load-leveling activities, power needs to be available for hours, and the related costs depend on the price of peak-load electricity. Investment costs for storage technologies are also highly variable, as these technologies are mostly in precommercial phases with little operating experience, except for pumped hydro technologies [4].

5.1.1 Typical Utility Operations

The electric power grid in the United States involves a complex network, where electricity is generated at power plants, transmitted to substations via 160,000 miles of transmission lines, and then distributed to various consumers (with differing energy demands) via distribution lines. While this structure has provided many benefits, its age, expansion, and high-demand areas have started to cause heavy burdens resulting in interruptions in electric service. In fact, the cost of power interruptions to electricity consumers in the United States has been estimated as approximately $80 billion each year, which is approximately one third of annual electricity costs [5]. Problems faced at the distribution level are generally the cause of these interruptions, and may be mitigated by distributed energy storage applications. Improving transmission and distribution (T&D) lines and building new generation plants is costly and time-consuming. Various energy storage technologies may be able to provide a reliable and cost-effective alternative approach. Energy storage technologies can respond to the grid faster than traditional generation sources, operate efficiently at partial load, and are able to vary discharge times depending on the application need. Therefore, these technologies are well-suited to handle momentary interruptions and meet peak demand without requiring major upgrades to grid infrastructure.

Electric grids are complex systems and require detailed management to ensure smooth operation and reliability. Independent system operators (ISOs) typically are responsible for monitoring and coordinating this complex structure, by attempting to maintain a real-time balance between the various facilities that generate the required electricity and the load (consumption of electricity). Demand and supply fluctuate throughout a day, and the ISO is responsible for managing power flows based on transmission constraints and for restarting the system in the event of a collapse [2].

Seasonal and daily patterns in electricity demand are driven by factors such as the need for heating, cooling, and lighting. To respond to this fluctuating demand, various types of power plants are built and operated by utilities. *Baseload* plants serve the purpose of meeting constant demand for electricity, and in the United States are typically provided by nuclear and coal-fired plants. Baseload plants are operated at full output as much as possible. While the output of baseload plants (especially coal) can vary, their high capital costs and low variable costs (due mainly to the fuel), make continuous operation more feasible. Moreover, especially in nuclear-based plants, rapid change in output needed to follow load is limited by technical factors. When variation in load is the concern, this variation is generally handled with *load following* or *cycling* plants, which typically are hydroelectric generators or plants fueled with natural gas or oil. Load following units are further categorized as *intermediate load* plants (for day-to-day variable demand) and *peaking* units (for meeting peak demand) [6].

Utilities also need to make additional plants available to meet unforeseen increases in demand, losses of conventional plants and transmission lines, or other unforeseen and unplanned incidents. This contingency service provided is referred to as *operating reserves* and includes meeting frequency regulation (the ability to respond to small, random fluctuations), load forecasting errors (the ability to respond to a greater or less than predicted change in demand), and contingencies (the ability to respond to an unscheduled power plant or transmission line outage). Frequency regulation and contingency reserves are collectively referred to as *ancillary services*, and require power generation units that can quickly change output. Due to the swift response required, plants that are online and "spinning"[1] are used. The need for spinning reserves increases costs and decreases efficiencies of power systems, due to several factors [6]:

- Plants providing spinning reserve must be operated at part load, so they potentially displace more economic units.
- Partial loading can reduce the efficiency of individual power plants.
- The reserve requirements increase the number of plants that are online at any time, which increases the capital and O&M costs.

5.1.2 Support through Energy Storage

In addition to energy arbitrage and operating reserves, there are several other services that energy storage technologies are capable of providing to the electric grid [6]:

- *Transmission and Distribution:* Energy storage is also capable of acting as an alternative or supplement to new T&D infrastructure. Distribution systems need to be sized to be able to meet peak demand, and as demand grows, new lines and substations need to be installed. However, these systems are generally only used for a few hours per year, as peak demand needs arise. The construction of new distribution lines is

1. Operating reserves met by spinning units are sometimes referred to as spinning reserves.

expensive (especially when they are not necessarily fully utilized all the time). This expense may be avoided or at least deferred by deploying distributed energy storage located near the load, thus allowing energy to be stored during off-peak periods when the distribution system is lightly loaded, and discharged during peak periods when the system may be overloaded.

- *Black-Start:* Black-start is the capability of supplying needed energy if a system failure has occurred, energy to help other units restart. Pumped hydro units have typically been used for this application.
- *Power Quality and Stability:* Energy storage may also be used to assist in a general class of services that include issues such as voltage spikes, sags, or momentary outages. These disturbances can limit the ability of utilities to transmit power and also affect the stability and reliability of the entire electric grid.
- *End-Use/Remote Applications:* In terms of end-use, energy storage can provide power for off-grid homes, but may also provide value when grid-tied through management of time-of-use rates, or demand charges in large commercial and industrial buildings, or in back-up power applications for increased reliability.

5.1.3 Integration of Renewable Energy into the Electric Grid

The inherent variability of renewable energy resources, and concerns regarding the reliability of an electric grid that derives a large fraction of its energy from these resources, along with the cost of integrating large amounts of variable generation into the electric grid have caused an increased focus on the deployment of energy storage as an essential component of future energy systems [6]. The operators of the electric grid are already faced with the challenges of trying to match the supply of electricity to the demand from consumers, on a day-to-day basis, even with conventional resources used to generate electricity. Thus, energy storage systems are a vital partner to renewable electricity generation, as the variable nature of renewable energy may level out, aiding with the integration of renewables with the electric grid and leading to reduced greenhouse gas (GHG) emissions. The extent of renewable electricity generation to date, along with expected future levels of generation, do not necessarily require energy storage. However, as renewable resources are used in growing quantities, storage technologies can facilitate large-scale integration of these resources with the grid by reducing related costs and managing generation variability [1].

Modeling and analysis was done by the California Energy Commission (CEC), in an attempt to understand the variability and system performance related to 20 and 33 percent renewable energy penetration into the electric grid. Results indicated that while system performance degrades (due mainly to both renewables and load ramping in the morning and evening) to a significant degree in the 20 percent renewables scenario, degradation reaches extreme levels in the 33 percent renewables scenario. These results indicate that increase in variability (as renewables gain an even bigger market share) will necessitate a considerable increase in ancillary services, especially frequency regulation [2]. Other analyses have demonstrated similar results. For example, the New York ISO conducted a study focused on wind generation, and concluded that for every 1,000 MW increase in installed wind generation (between the 4,250 MW and 8,000 MW penetration level) the regulation requirement would increase by 9 percent, which would mean that additional capacity is needed. [2]

The availability of wind and solar energy sources can display a large range of variance, from a few seconds to hours or even days, and these differences translate into different (power-and energy-related) energy storage requirements. Power-related requirements involve relatively low capacity, but, fast response; for changes that occur within seconds or over a period of a few hours. Energy-related requirements involve high capacity, but, slower response; for changes that extend over one or more days. However, energy storage requirements also extend continuously across the time spectrum, and many storage technologies span the two requirements [7].

5.1.4 Metrics for Evaluating Energy Storage Technologies

Even though each specific energy storage technology requires different specifications, there are three key factors that relate to all storage technologies: system economics, technical performance, and system design [8]:

- *System Economics:* This metric is the most important metric to the electric power industry. As consumers have become accustomed to continuous electricity service at affordable prices, the life-cycle cost of storage technologies to be deployed becomes even more important and critical to widespread adoption. The electricity industry generally fails to recognize the full benefits of energy storage, and view storage as a peak generation substitute, thus they believe that an energy storage technology must be competitive with the cost of currently available technologies used for peak electricity generation (such as gas turbines), while also providing increased efficiency and benefits able to offset capital, operating, and lifetime costs.

- *Technical Performance:* The technical performance of an energy storage technology impacts the overall system economics. Performance needs are generally application-specific, but include factors such as cycle life, energy density, response time, rate of charge/discharge, and efficiency.
- *System Design:* System design is related to the cost and technical performance of an energy storage system. This factor includes the storage device itself, the power conditioning and control systems that allow the system to communicate with the electric grid, and any other ancillary equipment needed for successful operation. The scalability of a system depends on these factors, and also on factors such as materials availability, automated manufacturing, and the complexity level of the system. The system design must also meet safety standards of the electric power industry and manage any health and safety risks to utility workers and the surrounding community.

Looking into more detail, energy storage technologies may be defined and evaluated further via the following characteristics [9]:

- *Energy Storage Capacity:* Energy storage capacity is defined as the quantity of energy (kWh or MWh) that can be stored at a given time. Since over-discharging some technologies (especially lead-acid batteries) results in shortening their useful lifetime, many storage technologies have limitations on how much of the stored energy may be used.
- *Charge and Discharge Rates:* Charge and discharge rates are measures of power (kW or MW) and designate the rate at which energy is either added (charge mode) or removed (discharge mode) from a storage system. In general, for most energy storage technologies these rates do not remain as constant value, but rather, they change with how much energy is stored and how long power has been continuously removed or added to storage. Charge rates are typically lower than discharge rates for most energy storage technologies, and these systems will commonly be described in terms of their discharge rates.
- *Lifetime:* Each energy storage technology has a limited lifetime and with age, performance is degraded. While some technologies measure lifetime based on "cycles," or how much they are charged and discharged, other technologies focus more on functionality over time, and yet others have lifetimes limited by total energy throughput.
- *Round-trip Efficiency:* Round-trip efficiency (expressed as a percentage) is the ratio of energy discharged from storage to the energy input into

storage, and is predicated on the loss of energy that is due to the fact that every storage technology requires more energy to charge than discharge. An energy storage system that is less efficient will require more electricity to store the same amount of electricity supplied, compared to a more efficient energy storage system. This translates into higher costs.

- *Capital Costs:* Capital costs for a specific energy storage system will vary based on factors such as system size, location of the system, local labor rates, market variability, intended use of the storage system, local climate, environmental considerations, and ease of access or transport. Installing an energy storage system will also add other costs, which are referred to as balance-of-system (BoS) costs, and include components such as safety equipment (e.g., fuses, current fault protection), inverters or rectifiers, system controllers, remote monitoring equipment, and supplemental sensors. The impact of the BoS on total system cost could be significant, ranging from 100 to 400 percent more than the costs of the storage technology itself. In comparing different energy storage technologies, the price per amount of power that the storage can deliver [$/kW or $/MW] and the cost per amount of energy capacity [$/kWhcap or $/MWhcap] are typically used as measures. The expected lifetime of the storage technology is also an important consideration, as systems with shorter lifetimes will require more frequent replacements, resulting in added costs. The entire lifetime cost of an energy storage system may be expressed as the capital cost divided by the total expected lifetime energy throughput [$/kWhlife].
- *Operating Costs:* To maintain peak performance, energy storage technologies need to be maintained. The operation of a system, plus maintenance, translates into yet another set of costs. A variety of factors have an effect on operation and maintenance (O&M) costs, such as how often the storage equipment is used, ambient temperatures, handling of the equipment, quality of installation, protection from overcharging/over discharging, the rate at which the equipment is cycled and the quality of the storage equipment. All of these factors are combined into an annual cost based on the size of the equipment, expressed in terms of $/kW × year.

Size and weight of storage devices are also important factors. Volume- and weight-based energy densities vary according to storage technology, as displayed in Figure 5.1, and these ranges reflect the differences among manufacturers, product models, as well as the impact of packaging [10].

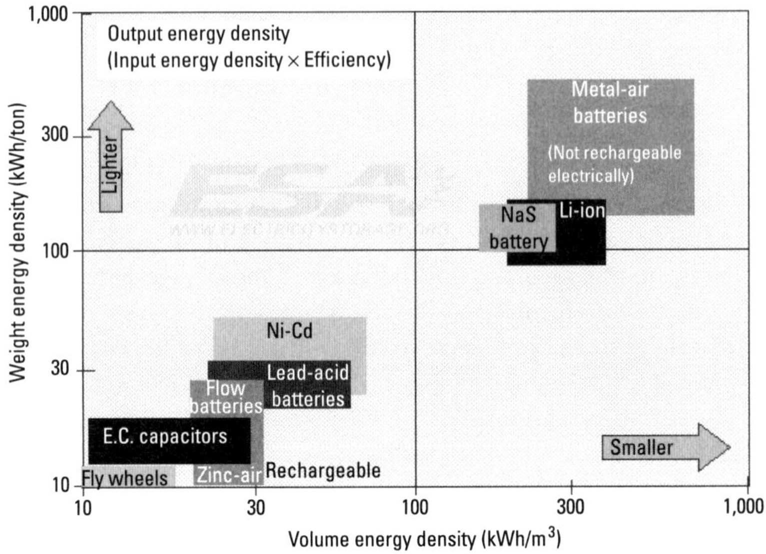

Figure 5.1 Comparison of volume- and weight-based energy densities of various energy storage technologies [10].

5.1.5 Energy Storage Technologies, Characteristics, and Applications

Energy storage applications are often divided into three categories [6, 10], based on the length of discharge: power quality, bridging power, and energy management. In *power quality* applications, the need for stored energy is very short (seconds or less) to assure continuity of quality power, and includes services such as frequency regulation. Technologies suitable for these applications include flywheels, capacitors, and superconducting magnetic energy storage (SMES). *Bridging power* applications require several minutes to about an hour, to assure continuity of service when switching from one source of energy generation to another. *Energy management* applications usually include continuous discharge of several hours or more. Storage technologies are used to decouple the timing of generation and consumption of electric energy, such as in load leveling, where energy is stored when costs are low and used as needed. Technologies for these applications include several battery types, pumped hydro, CAES, and thermal energy storage. Some storage technologies are cross-cutting and are capable of functioning in all functional categories, but most options would not be economical to be applied in all three functional categories.

Different types of energy storage technologies have performance characteristics and costs that make them suitable for particular types of applications. While some applications may emphasize fast response over duration, others may require longer durations, suggesting that no one technology fits all applications. Two characteristics are frequently used to characterize and compare different energy storage technologies: *rated power* (how much power a storage unit can provide, in MWs) and *discharge duration* (how long the power can be provided, in hours). There are cases of overlap in technology ratings, and even within a given technology type, the range of ratings can be large. Figure 5.2 compares these features for different energy storage systems [10].

Storage technologies may also be compared based on their capital costs, power rating ($/kW), and energy rating ($/kWh). Capital and operating costs are also important in determining whether a technology is suitable for a given application. For instance, even if a specific energy storage technology may seem suited technically for a given application, its costs might not be justified. Thus, both the technical capabilities and the financial viability of the storage technology are important in determining which technology is suitable for which application. The efficiency and cycle life (number of charge and discharge cycles that a storage device can provide before performance decreases) of an energy storage technology are also two important characteristics. These factors influence the lifetime costs of the storage technology [3].

The characteristics of an energy storage system mainly relate to its storage properties, operational properties, safety, location, construction time, and lead time. Storage properties involve factors such as energy density, output density, energy storage efficiency, scale of storage, and the time it takes to charge/discharge. The operational characteristics are tied to start/stop times, load re-

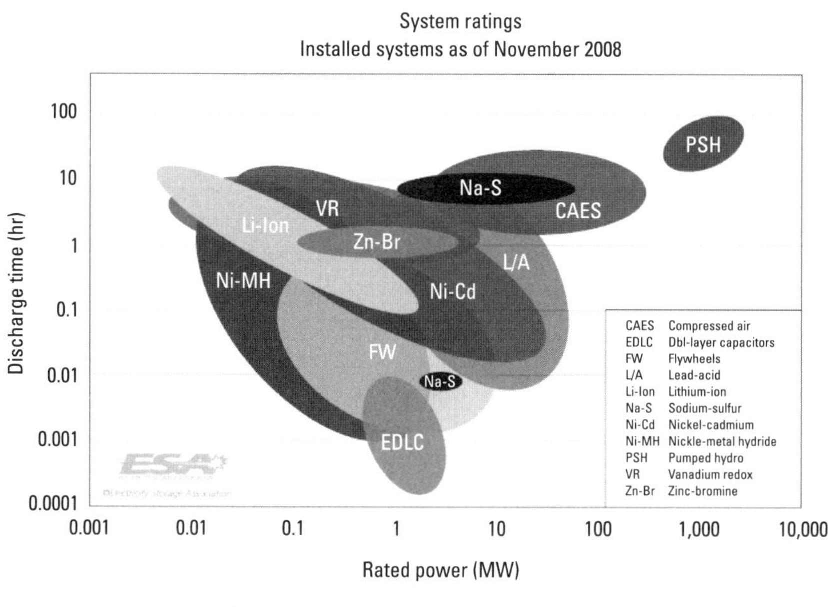

Figure 5.2 Comparison of rated power and discharge time of various energy storage technologies [10].

sponse, partial load features, lifetime, and reliability. While all these features are important, some may gain more importance depending on the use purpose.

The performance of a storage system may be defined by several key parameters [4]:

- *Energy storage capacity (kWh):* The amount of energy that can be stored in the system.
- *Charge and discharge rates (kW):* How fast energy can be charged or discharged. In general, the discharge rate can vary during operation and is often higher than the charge rate.
- *Response time (in seconds, minutes):* The time needed for the storage system to start providing energy on demand.
- *Lifetime:* The number of cycles, years or stored/provided energy (kWh).
- *Efficiency (or roundtrip efficiency, %):* The ratio of energy discharged by the system to the energy needed to charge it at each cycle. Also accounts for energy lost in the storage cycle.
- *Energy density (kWh/kg, kWh/m^3, Wh/l) and power density (kW/kg and kW/m^3):* These factors are especially important in applications where space is at a premium, as they denote the amount of energy or power provided per unit weight or volume.

The main advantages and disadvantages, as well as the feasibility of various energy storage technologies for high power and high energy applications are summarized in Table 5.1 [10].

5.2 Pumped Hydro

Pumped hydro systems are an old and mature technology; the first such system was built in 1892 [11]. This technology is capable of large capacity energy storage (hundreds to thousands of MWs), has a long lifetime (more than 40 years), is about 70 percent efficient, and has shorter response times (only requiring about a minute to go from zero to full output) [11]. Stored energy may be discharged over long periods of time (4 to 10 hours) [12].

Pumped hydro systems store electricity by using this electricity to run pumps, which pump water from a lower-level reservoir to a higher-level reservoir. While the low-cost electricity generated during periods of low demand operates the system, during periods of high electricity demand (and resulting higher prices), water flows back down to the lower-level reservoir, while turn-

Table 5.1
Comparison of Various Energy Storage Technologies [10]

Storage Technologies	Advantages	Disadvantages	Power Application	Energy Application
Pumped storage	High capacity, low cost	Special site requirement	Not reasonable or economical	Fully capable and reasonable
CAES	High capacity, low cost	Special site requirement, need gas fuel	Not reasonable or economical	Fully capable and reasonable
Flow batteries	High capacity, independent power and energy ratings	Low energy density	Reasonable for this application	Fully capable and reasonable
Metal-air	Very high energy density	Electric charging is difficult	Not reasonable or economical	Fully capable and reasonable
NaS	High power and energy densities, high efficiency	Production cost, safety concerns (addressed in design)	Fully capable and reasonable	Fully capable and reasonable
Li-ion	High power and energy densities, high efficiency	High production cost, requires special charging circuit	Fully capable and reasonable	Feasible but not quite practical or economical
Ni-Cd	High power and energy densities, efficiency		Fully capable and reasonable	Reasonable for this application
Other advanced batteries	High power and energy densities, high efficiency	High production cost	Fully capable and reasonable	Feasible but not quite practical or economical
Lead-acid	Low capital cost	Limited cycle life when deeply discharged	Fully capable and reasonable	Feasible but not quite practical or economical
Flywheels	High power	Low energy density	Fully capable and reasonable	Feasible but not quite practical or economical
SMES	High power	Low energy density, high production cost	Fully capable and reasonable	Not reasonable or economical
Electrochemical capacitors	Long cycle life, high efficiency	Low energy density	Fully capable and reasonable	Reasonable for this application

ing turbines to generate electricity. Figure 5.3 displays the typical layout of a pumped hydro system [11].

Pumped hydro systems may also be categorized as *mixed* or *simple* pumped hydro systems. A mixed system utilizes a large reservoir with a larger effective storage capacity, thus being capable of operating during high water seasons. In dry seasons, this system provides support in meeting the load. Typically 200 to 400 MW systems are used. On the other hand, a simple pumped hydro system has an artificial reservoir, thus requiring to be charged with water. In order to

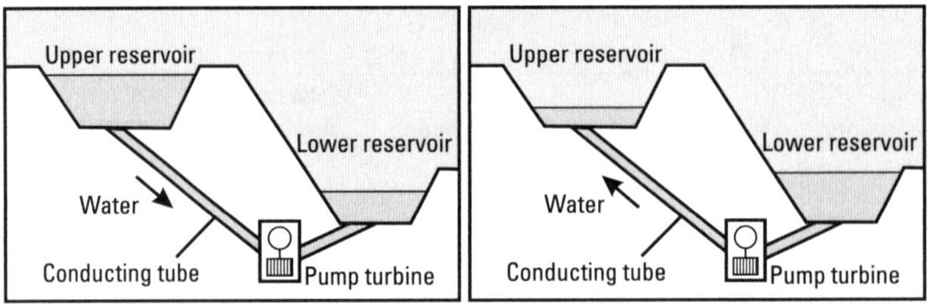

Figure 5.3 Typical pumped hydro system: Discharge (left); charge (right) [11].

respond to peak load in a short period, capacities of 1,000 MW to 2,000 MW are used [11].

However, pumped hydro systems face challenges due to location-specific constraints, as siting these systems is dependent on environmental, topographical, and geological conditions, which may be alleviated to some extent via the use of underground reservoir systems. While in conventional pumped hydro plants, both the upper and the lower reservoirs are constructed above the ground, systems that are based on underground reservoirs are of two types: freshwater-based or seawater-based. While both systems exploit artificial underground tunnels as the lower reservoir, a natural lake serves as the upper reservoir in a freshwater-based system and the ocean serves as the upper reservoir in a seawater-based system. Underground pumped hydro facilities reduce impacts on the above-ground environment, preserve the ecological system, allow for more choices in plant siting, and reduce transmission costs by allowing the installation of the system closer to electricity consumption [11].

The key characteristics of pumped hydro systems include [13]:

- *High power capacity:* The power capacity of pumped hydro plants are limited mainly by the maximum flow volume of the river (which is usually very high) and the number of turbines placed in the plant (which is customizable according to needs).
- *High energy capacity:* As large reservoirs within geographic features allows for the storage of a very large amount of water, a significant amount of energy may be stored.
- *Moderate response time:* The timescale of response for pumped hydro systems may be classified as moderate. It is limited by the variation of water flow and the inertia of the turbines, which while small, is still significant and limits response.
- *High round-trip efficiency:* The use of the pumps in pumped hydro systems is the main source of parasitic loss, but as these pumps have very high isentropic efficiencies (since the liquid is incompressible), overall

roundtrip efficiencies are relatively high, in the range of 80 percent to 85 percent.

- *Relatively low cost:* These energy storage systems can be considered relatively inexpensive, as they are usually created from retrofits of existing conventional hydropower plants and the components required are abundant.
- *Geographically limited:* The siting of pumped hydro plants (similar to the siting of conventional hydropower plants) needs to be on large rivers which have maximum capacity constraints, and is also limited by water use and flood control constraints.
- *Scale limited:* As pumped hydro systems are based at dams, they are large installations and thus not appropriate for small-scale applications. Rather, this technology is best suited for bulk, centralized energy management on the electric grid, for load-balancing in a wide range of applications such as renewable intermittency mitigation, load fluctuation management, and providing contingency power. Capital cost of new pumped hydro facilities is estimated to range between \$2,000/kW to \$4,000/kW. Dam and civil infrastructure accounts for 60 percent, pumps and turbines account for 15 percent, and other components and systems account for the remaining 25 percent. Costs are sensitive to the site, and cost reductions from learning are not really expected, as this is a relatively mature technology. The overall cost of pumped hydro energy storage is estimated at between \$50/MWh and \$150/MWh [4].

5.3 Compressed Air Energy Storage (CAES)

Compressed air energy storage (CAES) systems are also relatively mature. The first commercial CAES system built was a 290 MW unit built in Handorf, Germany in 1978, and the second commercial site was a 110 MW unit built in McIntosh (Alabama), United States in 1991 [14].

In compressed air energy storage (CAES) systems, electricity is used to inject air (at high pressure) into underground geologic formations, and when needed, high-pressure air is released from the underground and used in powering natural gas-fired turbines. Due to the high pressure involved, the turbines are able to use much less natural gas in generating electricity. CAES systems may be constructed in capacities of a few hundred MW and can be discharged over relatively long (4–24 hours) periods of time, thus making the systems suitable for load-leveling [1]. Compressing air during off-peak periods when energy prices are typically low allows electricity to be produced during peak periods at lower costs than if conventional gas turbines were used [14].

The typical components of a CAES system are a compressor/motor unit, gas turbine, and underground compressed air storage in salt caverns, as shown in Figure 5.4. Charging typically occurs at night, where the motor uses power to compress and store air underground. The compressed air is cooled via a cooler unit. Then during discharge, typically during the day, compressed air goes through a gas turbine, where the combusted gas is expanded through the turbine, which drives the generator and produces electric power [11].

Storage caverns are usually at depths of 500 to 800 meters, and the compressed air is stored at a pressure of 45–70 bars. The efficiency of a conventional CAES system is around 50 percent, but efficiencies may be improved with the use of an advanced gas turbine cycle. CAES units are able to come online within about 14 minutes. However, as natural gas is typically used to operate the gas turbine, the method results in carbon dioxide emissions. As it is most economical to locate CAES systems near salt rock deposits, siting opportunities are limited. However, above-ground storage in metal vessels or pipes is also an option, expanding the siting options for CAES systems [11].

Capitals costs for new CAES facilities range from \$800/kW to \$1,000/kW for capacities of 100 to 300 MW (if cheap underground storage sites are available) and \$1,500/kW to \$1,800/kW for smaller capacities of 10 to 20 MW. If the CAES systems utilize ground in-vessel storage, their costs are three to five times higher. The overall cost of CAES storage is \$150/MWh [4].

5.4 Flywheels

A flywheel stores electrical energy as rotating energy. During the charging phase, the flywheel rotates at high speeds, up to 30,000 to 40,000 RPM, and during

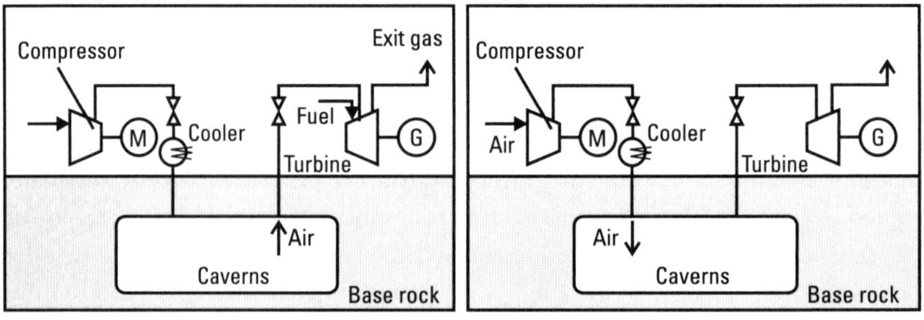

Figure 5.4 CAES system components: Daytime discharge (left), nighttime charge (right) [11].

discharging of electric power, the rotating speed of the flywheel decreases, while the spinning mass of the flywheel turns a generator that produces electricity. Thus, initially electric energy is transformed to rotating energy (during charging) and then rotating energy is transformed back to electric power (during discharging). Figure 5.5 displays the basic components of a flywheel [11].

The amount of energy that can be stored kinetically in a flywheel is a function of the cube of rotational speed, allowing higher speeds to provide higher energy storage densities [14]. Flywheels are able to provide kW- to MW-size power for a limited time (seconds to minutes) with very short response time (able to go into service in about 15 minutes), and thus are suitable for use for back-up power, frequency regulation, and wind power support in small grids [4; 14]. Empirical data have shown that a 1 MW flywheel system was about four times more effective in providing frequency regulation services than a slower fossil-based regulating generator [2].

Capital costs of flywheels range from less than \$1,000/kW for small, simple back-up power systems to \$4,000/kW for MW-size systems. The costs of potential very large (10 MW) systems is estimated to be approximately \$3,000/kW [4].

5.5 Batteries

Various types of large-scale rechargeable batteries may be used for stationary power storage systems, such as sodium sulfur (NaS), lithium ion (Li-ion), and flow batteries, in applications such as power quality and load-leveling. Moreover, as plug-in hybrid electric vehicles (PHEVs) become more prevalent, their connection to the electric grid will allow the use of their onboard batteries to

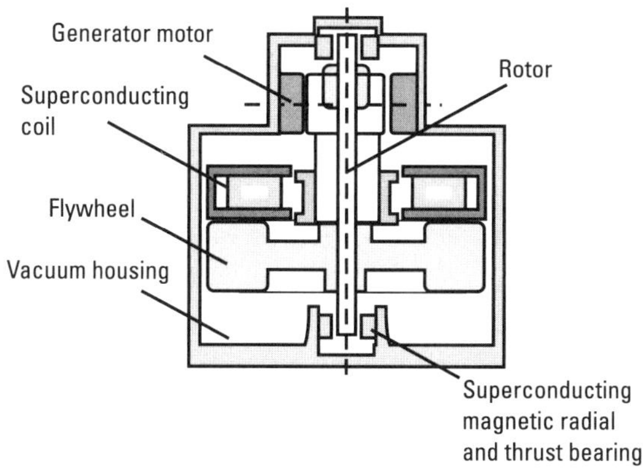

Figure 5.5 Components of a flywheel [11].

also be used for the storage of electricity [1]. Over the last 20 years, the development and use of batteries was focused on the consumer electronics and power tools sectors. However, within the last decade, related efforts have also diversified into the advanced design and use of batteries for the transportation sector. This in turn has opened-up the path to using batteries for electric grid energy storage applications. In order for batteries to be applied to the electric grid, power conversion systems that convert the direct current (DC) power supplied by batteries to the alternating current (AC) power that the electric grid operates on were needed. Development of these devices has advanced to allow various types of batteries to be used for grid support. The energy density of batteries has improved over time. Lead-acid batteries were initially introduced in the mid-19th century, with energy densities of 25–45 Wh/kg. In the first half of the 20th century, nickel-iron and nickel-cadmium batteries came into use, with energy densities ranging from 30 to 60 Wh/kg. The latter half of the 20th century saw developments in nickel-metal hydride batteries, increasing the energy density range to 50–75 Wh/kg. By the advent of the 21st century, lithium-ion batteries entered the market, with significantly increased battery performance, providing energy densities of 110–140 Wh/kg [14].

Lead-acid and NaS batteries have the most extensive track record in terms of large-scale energy storage. Lead-acid batteries were tested for utility peak shaving in the 1980s, but the economics at that time were not in favor to allow further deployment. However, improvements realized over time, such as three to four times the energy density with improved lifetimes over conventional lead acid batteries, as well as increased energy costs, have made the use of lead acid batteries more economical [7]. Lead-acid batteries, NaS batteries, and large Li-ion batteries are generally the technologies of choice for distributed storage applications such as residential/commercial PV systems. They may also be used for frequency regulation and back-up power systems. Among these technologies, Li-ion batteries provide the best energy density, cycle efficiency, durability, and lowest self-discharge. While smaller Li-ion batteries for portable devices are commercially available at relatively low prices, use of this technology for grid energy storage applications is still expensive. They cannot be scaled-up as easily and require enhanced safety and reliability measures [4].

Different battery types display different characteristics, but tend to have the following common properties [13]:

- *Relatively low energy capacity:* When compared to other energy storage systems, and to the scale of electricity utilized on the electric grid, the amount of energy that can be stored per unit mass of electrolyte in a battery is relatively low (0.14–0.46 MJ/kg), as chemical electrolytes are limited by the amount of charge that can be stored before reaching saturation. Significant amounts of electrolyte would be required to store large amounts of energy, which in turn would increase the battery size and cost considerably.

- *Relatively high power capacity:* Batteries are capable of providing large amounts of power, as the time required for the inherent chemical reactions is very short.
- *Fast response time:* The fast chemical reactions within a batteries also allow them to respond quickly (on the millisecond timescale) to changes in power.
- *High round-trip efficiency:* The charge/discharge cycle efficiencies of batteries are typically around 90 percent, depending on battery chemistry.
- *High cost:* The expensive materials and detailed processes required to manufacture batteries, combined with their low energy densities, result in high overall deployment costs relative to other energy storage systems, when used in large amounts. Thus, batteries are better suited for providing fast timescale services for large fluctuations in either electricity demand or renewable power output for shorter periods.

5.5.1 Lead-Acid Batteries

The lead-acid battery is the oldest and most mature of all battery technologies. It has been used in a wide variety of applications, providing for technology experience that leads to lower costs. In 1988, a 40 MWh lead-acid battery was installed in Southern California in a grid application to demonstrate the peak shaving capabilities of batteries. While the battery demonstrated the value of stored energy for the grid, its limited cycling capability made the overall economics of the system unacceptable. However, for back-up power applications, lead-acid batteries are still used in case of emergencies. The long operational life and lower costs they provide make them ideal for applications with low duty cycles [14]. However, lead-acid batteries face challenges involving cycle life, maintenance requirements, specific energy, and high-voltage operation [8].

5.5.2 Nickel-Cadmium Batteries

By the middle of the 20th century, the nickel-cadmium battery gained a reputation as a rugged, durable technology for storing energy, with good cycling capability and a broad discharge range. Nickel-cadmium batteries have been utilized in several back-up power applications, and are still being used for utility applications such as power ramp rate control for "smoothing" wind farm power variability in areas with weak power grids (such as island power systems) [14].

5.5.3 NaS Batteries

NaS batteries operate at high-temperatures (around 300°C), and consist of a liquid sulfur positive electrode and a liquid sodium negative electrode, separated by a solid ceramic electrolyte. During the discharge process, the positive sodium ions flow through the electrolyte and electrons flow in the external circuit of the battery, producing about 2 Volts. Charging is the opposite of this process, where the sodium polysulfides that were created release the positive sodium ions back through the electrolyte to recombine as elemental sodium. NaS batteries have high efficiency (around 89 percent) and are capable of six hours of discharge time. The technology behind NaS batteries was originally developed in the 1960s for the purpose of being used in electric cars, but was later abandoned for that application. The development of the technology for large-scale applications took place mainly in Japan [14].

The main challenge for current sodium-based batteries is that their costs are still high, and reducing these costs requires improvements in performance, reliability, and durability. These batteries also need to operate at high temperatures, so they require costly thermal management systems. Repeated freeze and thaw cycles reduce system cycle life significantly. The electrolyte structures used currently also limit the performance of sodium-based batteries and result in high production costs. The corrosive environment within these batteries prevents the use of cost-effective materials for packaging and degrades the materials and battery performance. The current size, weight, and high-temperature operation of sodium-based batteries also makes them difficult to transport, and this does not fit well with the needs of utilities as storage needs evolve with the grid [8].

5.5.4 Li-ion Batteries

Li-ion batteries have a positive electrode consisting of $LiCoO_2$ and negative electrode based on graphite. The basic operating principle involves the exchange of ions between the electrodes during the charging and discharging cycles. The voltages of Li-ion batteries are higher than other types of batteries, as a non-aqueous electrolyte is used, as opposed to aqueous electrolytes used in other battery types. Li-ion batteries also have higher energy densities, at 100–200 Wh/kg (300–400 kWh/m^3). A disadvantage of Li-ion batteries is their operating range being close to hazardous conditions. These batteries require a protection circuit to monitor charge and discharge, as over-charging or over-discharging may generate heat and pose a fire hazard [11].

To date, Li-ion batteries have been used more extensively for electric vehicles or small consumer electronics, which require high energy, power density, and specific energy due to the volume and weight constraints of these applications. However, for grid energy storage applications, cost and cycle life are more important, even though a high energy density is still desirable. Most of the existing Li-ion batteries used for stationary power applications have a short life cycle and high costs. It has been shown that 80 percent of the capital cost of Li-ion batteries is due to the relatively high cost of materials used. The capacity of current Li-ion batteries decreases after repeated cycling, which diminishes the efficiency. Moreover, there are issues related to heat management, safety, and reliability that need to be addressed. Li-ion batteries are unable to effectively operate at temperatures lower than −10°C and present a potential safety hazard at temperatures greater than 70°C. The considerable heat generated during operation require thermal management mechanisms [8].

Capital costs for Li-ion batteries for power applications are in the order of up to $2,500/kW. New materials, further development of the technology for this application, and overcharging protection and packaging to aid with safety issues all take part in this cost. An encouraging trend, however, is that the development of the technology for transportation applications is demonstrating relatively high learning rates, in the order of 30 percent, indicating that capital costs may decline at fast rate [4].

5.5.5 Flow Batteries

A flow battery is in essence a rechargeable battery that converts chemical energy to electricity through the reaction of two electrolytes flowing past a proton-exchange membrane. This structure makes it similar to a fuel cell, except that the reaction is reversible and the electrolytes are reused. Additional electrolyte is stored in external tanks and pumped through the cell to charge or discharge the battery, and the energy storage capacity is limited only by the size of the tanks. Thus, scale-up is relatively easy, and cost per unit of energy storage is usually lower than for nonflow batteries, which improves the attractiveness for larger applications. In addition, flow batteries have high efficiencies (over 75 percent) and long lifetimes (over 10,000 charge discharge cycles) [7]. An additional benefit is that flow batteries can operate at ambient temperatures, as opposed to the much higher temperatures of other battery types such as NaS batteries. Flow batteries are also relatively light in weight and have a longer operational life, as their main components are based on plastic components [14].

Vanadium redox flow batteries are a relatively new technology, where energy is stored chemically in different ionic forms of vanadium in a dilute sulfuric acid electrolyte. In various countries around the world, vanadium flow batteries of 800 kW to 1.5 MW are being demonstrated in different applications such as firming island grid capacity and grid peak shaving [7]. Vanadium redox flow batteries are able to provide power capacity from kW-size to MW-size and may be used to support wind power generation at a cost of $3,000/kW to $5,000/kW [4].

Zinc bromide flow batteries are another version of flow batteries, which are essentially regenerative fuel cells based on a reaction between zinc and bromide. While zinc bromide batteries use electrodes in the electrochemical reaction, the electrodes themselves do not take part in the reaction and thus, there is no electrode degradation with repeated cycling. Several peak shaving and island grid applications have also utilized zinc bromide systems, in the 200 to 500 kW range [7].

Flow batteries currently have the main challenges of their inability to fully meet the performance and economic requirements of the electric power industry. The unwanted cross-transport that can occur in these systems can result in losses in efficiency and can also contaminate electrolytes. Another factor that impacts the performance and reliability of these systems is the stability and durability of membranes and electrolytes at various temperatures in the presence of strong reduction and oxidation conditions. Hydraulic subsystems that are needed (valves, pipes, and seals) do not currently have the low cost, long life, robustness, and efficiency that are needed. In addition, flow batteries require real-time analysis and monitoring tools, while the electric power industry typically has a poor perception of the technology [8].

5.6 Electrochemical Capacitors (Supercapacitors)

Supercapacitors are similar to lithium-ion batteries. They store energy in the two series capacitors of the electric double layer (EDL), which is formed between each of the electrodes and the electrolyte ions. The large surface provided by these systems provides for much larger energy densities, compared to conventional electrolytic capacitors. While the electrodes usually are made with porous carbon, the electrolyte is either aqueous or organic. Aqueous capacitors have a lower energy density due to a lower cell voltage, but are less expensive and are also able to operate in a wider temperature range [14].

While supercapacitors have lower energy densities than lead-acid batteries, they are also capable of being cycled hundreds of thousands of times and have faster charge and discharge capabilities [14]. Supercapacitors also offer very short response time and may be used as instantaneous voltage compensators, or in combination with battery storage. Capital costs for supercapacitors are estimated to be between $1,500/kW and $2,500/kW [4].

5.7 Superconducting Magnetic Energy Storage (SMES)

Superconducting magnetic energy storage (SMES) systems use a coil with many windings of superconducting wire, where energy is stored and released along with the increases or decreases in the current flowing through the wire. The

SMES device needs to be refrigerated to maintain the superconducting properties of the wire materials, which results in added maintenance costs. SMES may also be used to improve power quality, as they provide short bursts of energy [1].

The capital costs of SMES systems are based mainly on the superconductors, and the operating costs are due mainly to the cryogenic cooling system. Capital costs range between $200,000 and $500,000 for SMES systems with energy storage capacities between 200 kWh and 1 MWh [4].

5.8 Hydrogen and Fuel Cells

An emerging pathway that can enable the storage of low-cost (off-peak) electricity is the conversion into hydrogen via electrolysis. Utilities have become more engaged in this process, especially for the storage of off-peak, cheap, and often curtailed wind power. Storage of electrical energy via hydrogen involves using electricity to split water into hydrogen and oxygen through an electrochemical process called electrolysis. The generated hydrogen may then be used to generate electricity via a hydrogen-powered combustion engine or a fuel cell for grid load management, or used as a transportation fuel for hydrogen fuel cell vehicles.

While the roundtrip efficiency involved in producing hydrogen via electrolysis with subsequent conversion back to electricity in a fuel cell is poor compared to other storage methods, the potential amount of energy stored is very large, and the ability to decouple energy storage from power capacity is an important advantage. An emerging and promising application involves using excess renewable power to generate hydrogen through electrolysis, and then using the hydrogen as a vehicle fuel. This method may create a business case for renewable power providers if there is a strong vehicle fuel demand for hydrogen, and avoids the inefficiencies and equipment costs of converting the hydrogen back to electricity. The challenges involved in siting electrical transmission infrastructure for directing remote renewable power to urban areas may also be overcome, while a new value-added product is being created [13].

5.9 Thermal Energy Storage

Concentrating solar power (CSP) plants provide thermal energy storage by storing the solar energy in the form of heat in a synthetic oil or molten salt, extending power production for one to ten hours past sunset [1]. The issue of efficiency of thermal energy storage systems is not as simple as other storage technologies. In any energy storage technology, efficiency loss is an important issue. Electricity is considered a "high quality" source of energy, and when it goes through the process of being stored and discharged back, considerable losses are realized. Thermal energy, on the other hand, is a much lower quality of energy, but can be stored with much higher efficiency. Thermal energy is stored before conversion to electricity in a CSP plant, which results in the round-trip efficiency of CSP thermal storage to be close to 100 percent. However, the CSP thermal storage pathway is limited to storing thermal energy produced from the solar field, as opposed to other storage technologies that can store electricity produced from any source [6].

There are also thermal energy storage devices that can be deployed at the residential and commercial level. These include modular ice storage systems (generates ice during off-peak power periods to power air-conditioning systems during the peak afternoon load times), or modular heat storage systems (captures electric power during off-peak periods to use to store heat in a ceramic heat sink to be dispatched during higher peak periods) [14].

5.10 Impacts

Each storage technology has costs and benefits that need to be evaluated and utilized in evaluating which technology best fits the application.

5.10.1 Costs of Energy Storage Technologies

Determining the costs of energy storage systems in detail is a more complex process, as the costs of a specific technology are influenced to a great extent by the application of thetechnology. Estimating the total installed cost of a given energy storage technology depends on knowledge of a variety of key factors: *size*, which varies on two dimensions (power: how much electricity can be discharged at one time; and energy: how many hours can be discharged continuously); *efficiency* (how many useable kWh can be discharged compared to the amount charged); and *frequency* (how often and deeply the system is discharged) [12].

Capital and operation and maintenance (O&M) costs are the main cost components of an energy storage system. O&M costs include the cost of buying the energy used to charge the system, fixed costs which are not dependent on how often the system is used, and variable costs (mainly replacement costs).A metric that is commonly used to estimate life-cycle costs by incorporating these two cost components is $/kW-year, which denotes how much a kW of capacity costs to own and operate for one year [12].

5.10.2 Benefits of Energy Storage Technologies

Energy storage technologies can provide various benefits across the energy delivery value chain from generation to transmission and distribution, to end-users.

The multiple services provided by energy storage systems results in varying benefits, depending on the particular application of storage. Moreover, the value of an energy storage system is often divided between the owner of the system, utility shareholders, and utility ratepayers. This partition makes it difficult for one set of stakeholders to capture enough of the offered value to outweigh the technology's costs. The value provided, or benefits, of energy systems may in general be categorized as *economic* and *operational* benefits [12].

Economic Benefits

- *Savings related to shifting demand to off-peak times:* Energy storage enables customers to alter when they draw power from the grid to meet their demand. Opportunities are created to store energy when the cost of energy is low and discharge stored energy for use when the cost of energy is high, thus shifting load. The value of this shift in load varies based on the customers' load shape, tariff, and the timing and frequency of when the load is shifted. If the difference between peak and off-peak electricity rates is large enough, the utilization of energy storage systems can reduce energy charges.
- *Profits from selling energy storage services:* If there are market rules in place to allow energy storage technology owners to sell into ancillary services markets or wholesale energy markets, profits may be gained from these services.
- *Lower future energy storage costs:* The energy storage market is currently an emerging one, but as it matures, costs will decrease as a result of learning-by-doing, economies of scale, and R&D efforts. At the same time, increases in demand will encourage technology manufacturers, integrators and installers to become more efficient, which in turn would result in further cost reductions.
- *Local employment and other economic growth:* The locations where manufacturing facilities for energy storage technologies are built, as well as the sites where they are deployed, are poised to become areas of job growth, boosting the local economy,

Operational Benefits

- *Improved power quality:* Energy storage systems can eliminate power quality inconsistencies that may be faced by some commercial and industrial customers, thus saving them from the potential economic and other damage.

- *Reliable and cleaner back-up power:* During periods where power is not available from the utility system, energy storage technologies can provide customers with electricity for a period of several hours.
- *Reduced need for peak generation capacity:* By providing the ability of store energy off-peak and discharge on-peak, storage provides an alternative to the construction and operation of fossil-based conventional generators. As peak demand continues to increase and mitigating carbon emissions becomes more of an expense, the value of the avoided cost of peak generation capacity will also continue to increase.
- *More efficient use of renewable and other off-peak generation:* When renewable power sources (especially the case with wind energy) are generated at quantities much higher than demand, they often are curtailed. If energy storage technologies are utilized, rather than forcing renewable generators to curtail off-peak production, excess generation may be stored and used during high demand times, thus adding more value, instead of wasting a valuable product.
- *Reduced need for transmission and distribution capacity upgrades:* As energy storage technologies are able to shift demand off-peak, this means that they are also able to delay the need for upgrades to the T&D system which would have been needed to accommodate growth. The value of this T&D upgrade deferral differs by location, population density of the area, terrain, geology, weather, and the type and amount of T&D equipment involved.
- *Transmission support and congestion relief:* Energy storage technologies may also be beneficial in improving the T&D system's performance by easing problems related to issues such as voltage sag and unstable voltage, while also avoiding transmission congestion by discharging in congested areas at times of peak demand.
- *Lower GHGs and other emissions:* Energy storage allows reductions in emissions by shifting on-peak energy use to off-peak periods. As renewables increase as a percentage of the off-peak power mix, the emissions benefits of energy storage will continue to grow. Energy storage also serves as a lower-emissions alternative for providing ancillary services.

5.10.3 Comparative Evaluations

Research and detailed evaluations have been conducted by the U.S. National Renewable Energy Laboratory (NREL), to compare hydrogen-based energy storage to other energy storage technologies. Figure 5.6 reveals the results in

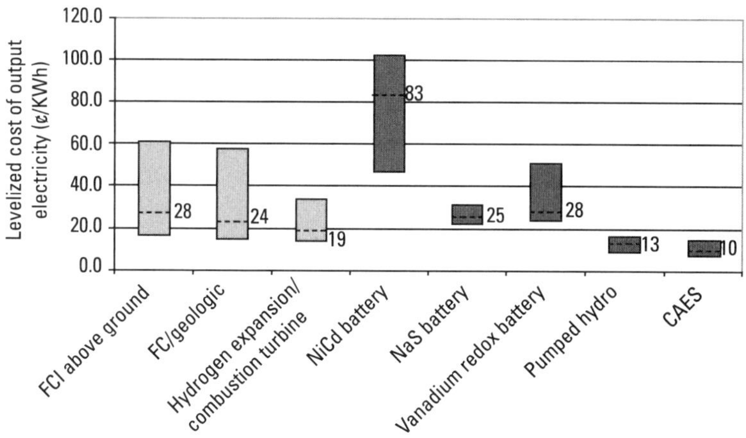

Figure 5.6 Hydrogen-based energy storage vs. other energy storage technologies [15].

terms of levelized cost of electricity for the various energy storage technologies investigate [15].

This analysis has demonstrated that, if cost reductions in hydrogen technologies are achieved, hydrogen could become competitive with batteries, and under specific circumstances, advanced hydrogen storage systems may prove to be cost competitive with pumped hydro and CAES systems [15].

The energy storage scenario used by NREL in this evaluation was a simple energy arbitrage scenario for a mid-sized energy storage system with a 300 MWh storage capacity charged during off-peak hours (18 hours per day on weekdays and all day on weekends) and discharged at 50 MW per six peak hours on weekdays. Two scenarios for producing hydrogen in excess of what would be required for the storage system were evaluated. In one scenario, 1,400 kg H_2/day were produced for the vehicle market, while in the other scenario enough excess hydrogen was produced to feed 500 kg H_2/hour into a hydrogen pipeline [15].

Additionally, the Argonne National Laboratory (ANL) in the United States investigated the potential energy- and emissions-related benefits of integrating hydrogen storage with renewable power generation, as well as the use of other energy storage systems, including pumped hydro, CAES, and vanadium-redox batteries [16].

The GHG emissions benefits of the different energy storage systems studied by ANL were evaluated for four different states of the United States, and per kWh into storage. The impact of roundtrip efficiency of the various energy storage systems is apparent in that greater GHG benefits are observed for systems having higher round-trip efficiencies. At the same time, the benefits realized are higher for the states where electricity production is mainly based on fossil resources, as these storage systems displace more carbon-intensive generation.

In the system using hydrogen for storage, oxygen is created as a byproduct. The credit received for it partially compensates for the system's low roundtrip efficiency and improves its competitiveness with the other storage systems [16].

5.11 Market Status

The energy storage market is developing rather quickly and is projected to increase over 20-fold, from $1.5 billion in 2010 to about $35 billion in 2020. However, many electricity storage technologies are still evolving and need policy support for further commercial deployment. The electricity storage capacity in today's grids is modest. More electricity storage could help ensure grid balance and reduce the need for costly peak-load capacity additions. On a global scale, simulations of 15–30 percent net wind power variations have shown a need for storage capacity from 190 GW to 300 GW by 2050 [4].

Pumped hydro and CAES technologies have moderate expansion potential, and are limited by their need for suitable installation sites. Energy storage via pumped hydro is reaching saturation in the European markets, while it will not be applicable in certain markets where inadequacy of water resources is an issue (such as the Middle East or certain parts of the Asia Pacific region) [17].

Installed energy storage technologies worldwide totaled over 125 GW in 2011. Pumped hydro was the main energy storage technology deployed globally, accounting for about 98 percent of installed storage technologies. Japan and Western Europe have led the deployment of pumped hydro technology. Approximately 2 percent of nonpumped hydro technologies have been dominated by thermal energy storage, batteries, and CAES [18]. In the United States, energy storage in the grid totaled over 23 GW in 2011, with 96 percent of this capacity provided by existing pumped hydro systems.

During the second quarter of 2012, the majority (close to 40 percent) of global installed energy storage capacity was in the Asia Pacific region, followed by North America (25 percent) and Western Europe (22 percent) [19].

Over the next decade, it is forecast that the market for advanced batteries in utility-scale applications will be modest. Growth is expected to be slow in the near term, as data is gathered on the operating characteristics, operations and maintenance requirements, and life-cycle costs of projects that are executed. However, during the latter part of the next decade, as the value proposition of these technologies is developed, a more robust penetration of advanced batteries is expected in the utility sector. It is anticipated that providing ancillary services will be the most attractive application for advanced batteries, while over the next decade, functions such as integration of renewables into the grid and T&D deferral are more likely. Each type of advanced battery provides different advantages, and comes with a related cost. Thus, different batteries may be more suitable for different grid applications, but more empirical data are still needed to fully confirm which battery suits which application best. Even with these unknowns and uncertain value propositions, advanced battery technology is being deployed and tested by various utilities around the world. In these initial development stages, though the overall installed capacity is low, the market for advanced batteries is displaying a considerable growth trend, with close to 16,000 MW expected to be installed by the end of the following decade [20].

Globally, the greatest potential for market development in energy storage technology lies in Asia. The Asia Pacific region currently has the largest base of energy storage, with 60 GW of installed capacity [21]. The diverse applications and benefits of energy storage technologies are garnering the attention of several nations in the Asia Pacific region. The region displays diversities in grid development levels, economic conditions, and demographics, resulting in differing requirements and demands for energy storage technologies in each specific country. A trend that is common to all nations in the region is that as renewable energy capacity and electricity consumption rise, the number of energy storage projects deployed in the region is expected to rise as well [22].

Japan is expected to place more emphasis on energy storage, especially as nuclear energy is projected to drop in share as the country continues to shut down nuclear power plants in danger of being damaged by earthquakes. Japan implemented feed-in tariffs in July 2012, where utilities are required to buy electricity from renewable sources for up to 20 years, at rates of 42 yen ($0.53) per kilowatt-hour for solar-generated electricity. This tariff is double that applied in Germany (where its solar industry tripled in output in less than a decade due to this support), and is more than triple the tariff paid in China. The Japanese government thus expects capacity from renewable energy to surge from 19,500 megawatts now to 22,000 megawatts in just nine months, and to handle this, it is estimated that by 2020, over 25 GW of storage will be needed for every 100 GWh of energy [21].

In Japan, Australia, and South Korea, as efficiency on the grid gains importance, attention will be paid to maintaining the balance between load and generation. Another trend is that developed countries in the region are actively preparing for community and residential energy storage systems, and various demonstration projects focused on integrating renewables are being deployed to deliver multiple applications to the end user [22].

The government and market stakeholders in China are entering an era of focus on R&D for bulk storage-level system designs for energy storage technologies, while also designing comprehensive legislation and government support. China has also strengthened its commitment to renewable wind energy and has increased its investments in solar power. This commitment to renewables, combined with a rise in electricity demand form the residential and commercial sectors, could drive the country to place more emphasis on energy storage technologies [22].

Most of the investment and attention related to energy storage in the Asia Pacific region during the decade between 2012 and 2022 is forecast to be focused on the advanced battery markets. The penetration of advanced batteries in the Asia Pacific region is expected to grow, with Li-ion forecast to take a higher share of the installed capacity in the region by 2022. However, the cumulative capacity of advanced batteries in Asia Pacific by 2022 is still expected to be somewhat modest, due to the slowness of adopting innovation in the utility sector. A cumulative installed capacity of over 25,082 MW is projected by 2022, with nearly 6,400 MW of capacity added that year [22].

5.12 Challenges and Potential Solutions

Various challenges, which are generally interrelated, currently exist on the path to more widespread commercial deployment of energy storage technologies [1, 5]:

- *High Capital Costs:* Compared with natural gas generators which provide similar grid support services, the capital costs of current energy storage technologies are still high.
- *Demonstration Project Needs:* In order for utility operators to increase their confidence in various energy storage technologies some large-scale demonstration projects are needed. More real-world performance data is required from large-scale demonstrations. These demonstrations will allow better definition of specific storage applications, while assuring stakeholders and regulators of the benefits provided. Large-scale demonstrations can also help leverage related manufacturing capabilities and infrastructure.
- *Transmission Planning:* The transmission structure only takes into account the location of demand centers and generation facilities, not remote energy storage facilities such as pumped hydro or CAES.
- *Regulations:* Energy storage technologies are considered by federal and state regulations as electric generation technologies, as opposed to investments in transmission capacity, thus avoiding transmission and distribution companies owning these systems. Furthermore, most incentives only consider renewables and exclude energy storage systems, even though these technologies can help facilitate higher penetration of renewable resources.
- *Insufficient Technical Progress:* The current high cost of various storage technologies, as well as their complexity present barriers to large-scale

production and deployment of these technologies. Moreover, some storage technologies, such as high-speed flywheels, electrochemical capacitors, and Li-ion batteries, have limited storage durations and energy capacities falling short of meeting the current needs of the electric power industry. Efficiencies and lifetimes of these various energy storage technologies also have limitations that create concerns for the electric power industry. Also, the required control systems and power electronics are currently not advanced enough to ensure integration into the grid reliably and securely.

- *Technology Risks:* In general regulated utilities are risk averse to investing in new technologies with their related high initial costs. Also, the energy industry invests a much smaller portion of their resources in R&D, compared to other industries, which further limits the extent of progress in these technologies.
- *Incomplete Electricity Markets:* Markets and transparent pricing structures have not yet been fully set in most regions of the United States for all of the various services that energy storage technologies are capable of providing. Since the current electric power industry classifies the market segments of generation, transmission, and distribution, it is difficult to fit energy storage into a classification and to measure its value compared to traditional infrastructure since storage is capable of supporting all functions of the electric grid system. As the pricing mechanism of energy storage is also dependent on its classification, without proper pricing, it becomes challenging to confirm that stakeholders will be compensated for the benefits of energy storage.
- *Weak Stakeholder Understanding:* While energy storage technologies have a significant potential to enhance the electric grid, their benefits are not yet fully understood by stakeholders. Utilities, grid operators, and energy storage developers are not adequately aware of the value of energy storage technologies and their applications. The lack of industry understanding also affects decisions of regulators in a negative manner, until they are able to receive adequate evidence of viable business models for energy storage, which need to account for financial, technical, and environmental value propositions. Moreover, if industry and regulators lack a clear understanding of the benefits of energy storage, it will naturally be difficult to influence the general public about these benefits, which is an important component especially in the siting of energy storage installations.

Regulation related to energy storage for grid applications has several layers of complication. Utility-scale energy storage, except for pumped hydro, is not yet widespread, and is only being used in pilot projects or site-specific projects. Therefore, utilities have uncertainties related to how investment in energy storage technologies will be treated, how costs will be recovered, or whether energy storage technologies will be allowed in a particular regulatory environment. As energy storage applications are capable of providing functions related to both generation and transmission, regulatory treatment of investments in energy storage is further complicated. In the example of a utility storing electricity generated during low-cost (off-peak) periods, then providing that stored energy back to the grid during high-cost (peak) periods, two different views can be taken. From a regulator's perspective, the energy provided from storage may be translated as generation. However, providing this stored energy also enables reduced transmission congestion, provides voltage support at a time of peak use, and provides other ancillary services that support transmission functions. The current system also does not fully credit the value of storage across the entire utility value chain. As generation, transmission, and distribution have historically been viewed as independent components of the grid system, cost recovery for grid-level energy storage investments is challenging [7].

5.13 Emerging Opportunities

Energy storage technologies are able to act to serve a multitude of services across several sectors, as detailed in the following sections.

5.13.1 General Emerging Trends in the Energy Storage Market

There are several key emerging trends in the energy storage arena [23]:

- *Large companies are entering the energy storage market:* Energy storage technologies have traditionally been viewed as the play ground of start-up companies and the venture capital investors that support them. Related efforts had been focused on markets with lower-rated energy requirements (in the kWs), mainly the consumer electronics and electric vehicle markets. However, recent trends display interest and investment in storage technologies that could scale to the grid level. Bigger corporations have noticed this trend and have started supporting these efforts financially.
- *While there is motivation, projects are sometimes small:* It is estimated that the grid storage market value is over \$15 billion today, on a global scale. Due to this relatively large market size, it is often assumed that a storage

solution must start at MW scale. However, that is not necessarily the case. Storage can provide value for applications that can require many megawatts of storage at a particular site (such as load shifting), but also for smaller-scale uses at the community or substation level. Moreover, as any investor would want to demonstrate the technical feasibility and economic benefits of a technology before deploying it further, initial small-scale deployments are typically used. Partnering vendors with large IPPs, companies, or utilities at an early stage is also an important opportunity for all involved. In an emerging market, such an approach becomes even more important, as customers may not fully understand what they want and vendors may not know whether customer requirements are technically or economically feasible.

- *Energy storage technologies are still expensive:* Despite success stories and development trends, most current energy storage solutions still do not fully meet market needs. Capital expenditure required for storage is still high, generally about double the amount for storage to be competitive on an unsubsidized basis with other nonstorage solutions that can accomplish at least some of what storage is capable of accomplishing (such as demand response or gas-fired peaking units). Additionally, customers have observed that vendors often underestimate installed costs, which would include costs related to interconnection, power electronics, and other operations and maintenance costs. The energy storage market remains a promising market, but significant cost reduction and technology improvements are needed before technologies can be incorporated at scale.

5.13.2 Renewable Power Systems for Remote Communities

Globally, there are numerous remote communities that have unstable grid connectivity, or completely lack this connection, and generate most or all of their electricity with diesel generators, and importing fossil resources. While capital cost is not much of an issue with these diesel generators, there are several factors working against their use in these isolated communities. Diesel generators have high operating costs, due to the low efficiencies of these units, difficulties and high costs relating to transport of the diesel fuel to remote locations, and variable fuel prices. Moreover, diesel generators emit GHGs. Therefore, remote communities have interest in better utilizing their own local, renewable resources, which are especially attractive for these locations, as they are capable of bringing clean energy to locations that would not normally be served through a grid extension in an economical manner.

Off-peak renewable energy may be stored in the form of hydrogen (produced through renewable electricity applied to an electrolyzer), which in turn may be used to generate power during peak times through a fuel cell. Renewable energy systems do typically have higher capital costs. However, under the circumstances of power generation for remote communities, they become favorable due to their lower life-cycle costs compared to diesel generators (as operating costs are lower for renewables). Use of hydrogen fuel cells for renewable energy storage allows for shorter payback periods relative to the deployment of diesel generators, are achievable. Hydrogen fuel cell systems also are more efficient (resulting in lower GHG emissions), are easily scalable to meet power requirements, and are capable of maintaining high efficiencies at all power levels [24].

An economic analysis was conducted to compare the capital cost of a hydrogen energy storage system (including the electrolyzers, compressors, storage tanks, and fuel cells) to a NaS energy storage system, assuming an off-peak electricity price at $0.04/kWh. Results of this analysis indicate that at around nine hours of energy storage, the hydrogen storage systems is able to provide both a lower capital cost and a lower levelized cost of energy compared to the NaS systems. This primarily results from the fact that additional (and expensive) batteries need to be added to increase the energy storage duration of NaS systems, while the hydrogen systems only require additional storage tanks (which are relatively inexpensive). Additional electrolyzers and fuel cells, typically the highest cost components of such a system, were not found in this analysis to be needed to increase the energy storage duration, but might need to be considered in other situations [24].

5.13.3 Vehicle-to-Grid Applications

An emerging storage-related application involves the cross-cutting area between advanced transportation and stationary applications. Depending on the growth and success of plug-in hybrid electric vehicles (PHEVs), an opportunity arises for using the onboard energy storage of PHEVs as distributed energy storage that would be available to the electric grid while the vehicles are plugged-in and recharging. Studies have indicated that two million electric vehicles with a 33-mile electric range could provide enough grid stabilization to accommodate 10 GW of wind energy [4]. PHEVs could provide discharging capability for the grid to improve grid utilization, to level demand, and to improve grid reliability. However, a challenge to this type of a structure would be how PHEV usage will interact with high levels of renewable energy generation capacity, when solar and wind power vary diurnally. If the PHEV charging load matches peak renewable energy production (such as wind energy generation during the off-peak hours of the night) then the PHEV load and renewable source will be well matched temporally. However, if the opposite is true, then the resulting mismatch may be problematic, and deployment of energy storage technologies will take on an even more important role. Smart grid technologies that enable time-of-use pricing could incentivize consumers to match their vehicle charging with times of higher renewable generating capacity.

References

[1] Pew Center on Global Climate Change, *Climate TechBook: Electric Energy Storage*, May 2009.

[2] Lin, J., G. Damato, and P. Hand, "Energy Storage--A Cheaper, Faster and Cleaner Alternative to Conventional Frequency Regulation," Report prepared by Strategen Consulting for the California Energy Storage Alliance, February 16, 2011.

[3] KEMA, Inc., "Market Evaluation for Energy Storage in the United States," Paper prepared for the Copper Development Association, Inc., 2012.

[4] International Renewable Energy Agency, *Electricity Storage: Technology Brief*, April 2012.

[5] Nexight Group, *Electric Power Industry Needs for Grid-Scale Storage Applications*, December 2010.

[6] Denholm, P., et al., *The Role of Energy Storage with Renewable Electricity Generation, (NREL/TP-6A2-47187)*, Golden CO: National Renewable Energy Laboratory, January 2010.

[7] American Physical Society, "Integrating Renewable Electricity on the Grid," Report by the APS Panel on Public Affairs-Committee on Energy and Environment, November 2010.

[8] Nexight Group, *Advanced Materials and Devices for Stationary Electrical Energy Storage Applications*, December 2010.

[9] International Renewable Energy Agency, *Electricity Storage and Renewables for Island Power: A Guide for Decision Makers*, May 2012.

[10] Website: Electricity Storage Association, Technology Comparison, a http://www.electricitystorage.org/technology/storage_technologies/technology_comparison.

[11] Inage, S-I., "Prospects for Large-Scale Energy Storage in Decarbonised Power Grids," (International Energy Agency Working Paper), 2009.

[12] Deal, M., et al., "Electric Energy Storage: An Assessment of Potential Barriers and Opportunities," California Public Utilities Commission, Policy and Planning Division Staff White Paper, July 9, 2010.

[13] National Petroleum Council, "The Interaction Between Plug-in Electric Vehicles, Distributed Generation, and Renewable Power," Working Document of the NPC Future Transportation Fuels Study, August 1, 2012.

[14] Roberts, B., "Capturing Grid Power," *IEEE Power and Energy Magazine*, July/August 2009, pp. 32-41.

[15] Steward, D., et al., *Cost Analysis Highlights Hydrogen's Potential for Electrical Energy Storage, National Renewable Energy Laboratory-Hydrogen and Fuel Cell Technical Highlights (NREL/FS-5600-48437)*, December 2010.

[16] Elgowainy, A., "Emissions Analysis of Electricity Storage with Hydrogen, DOE Hydrogen and Fuel Cells Program FY 2011 Annual Progress Report," November 2011.

[17] Dehamna, A., and E. Bloom, "Energy Storage on the Grid," Research Report, 3Q 2011, Pike Research, 2011.

[18] Electricity Advisory Committee, *Energy Storage Activities in the United States Electricity Grid*, May 2011.

[19] Gibson, B. E., K-A. Adamson, "Energy Storage Tracker," Research Report, 2Q 2012, Pike Research, 2012.

[20] Gibson, B. E., K-A. Adamson, "Advanced Batteries for Utility-Scale Energy Storage Applications," (Research Report, 3Q 2012), Pike Research, 2012.

[21] Simpkins, J., "A Brand-New Market Set to Grow 4,000% in the Next Decade," Wall Street Daily, July 11, 2012.

[22] Bae, A., and A. Dehamna, "Energy Storage in Asia Pacific," Research Report, 4Q 2012, Pike Research, 2012.

[23] Gould, J., "What We Learned in Washington, D.C.: Notes from U.S. Department of Energy's Storage Conference, Cleantech Group," Research Notes: Energy Storage, December 2010.

[24] Glandt, J. D., "A Discussion of PEM Fuel Cell Systems and Distributed Generation," BALLARD White Paper, May 2011.

6

Toward Sustainability: Putting the Pieces Together

While there is growing consciousness about the impact we are having on our planet due to the unsustainable patterns of energy use and living, change toward a more sustainable path will not be immediate or easy due to several complicating factors. Progress toward is still slow; slower than what is needed. On an individual level, there are great advances related to renewable energy. However, taking a step back and looking at the big picture, the sustainable portion of our energy consumption is still relatively small. Developed nations may have realized reductions of growth in energy demand, but developing nations, with much larger populations, are on the same unsustainable path, with potential for more damaging outcomes as the planet is already under stress. Another complicating factor is that accomplishing change is harder due to the human factor; things like uncertainty, social norms, and vested interests. Uncertainties relating to different energy supply options and their impacts, when combined with economic and political interests, along with basic resistance to change, result in major barriers that slow progress toward a more sustainable path. In addition to these complicating factors, *time is too short!* We are already past the point where we can mull over options at a leisurely pace. The time to act was *yesterday*, and as more time passes, it also becomes much harder to reverse negative impacts and develop solutions in time.

Looking at the whole picture, it can be seen that humans have come to full circle regarding their harvesting patterns of the resources required for energy production, but with some variances. Early humans looked at the sun, wind, and water for the power to improve their lives. Then there was the realization that solar energy had been concentrated and stored underground in the form of fossils, and human ingenuity found a way to take advantage of this previously hidden resource. As time passed, this new resource became the mainstay of human activity and societies. However, these developments did not come without a price to humans, the ecosystem, and the planet as a whole. Mankind realized that its insatiable appetite for energy and natural resources had started to leave dents in the ecosystem it depended on. Once again we started reflecting back on the sun, and the wind, but now with a different perspective, equipped with better knowledge and technical capabilities. Centuries of developments in science and technology had equipped mankind with effective tools, techniques, and insights. These tools, combined with lessons learned from the past and human creativity hold the potential for new solutions resulting in more sustainable energy development and use. However, there's also a twist in the plot: now we're much more crowded, have an increasing appetite for energy, and have already left behind a large footprint. Most importantly there's already an established, working, cheap energy-related infrastructure that mankind has grown accustomed to. Change regarding any subject, at any level, is always hard. But, those who take the risks and put in the required effort are those who reap the greatest rewards.

The pursuit of solving our energy dilemma cannot be looked at as a one-sided challenge. We have already put our planet under stress through damages to all components of the ecosystem. It has been estimated that we already have gone past certain limits. While our ecosystem has a remarkable capability to replenish itself, our actions have brought us to a point where it takes the Earth a year and a half to regenerate what has been used-up in a year [1]. Thus, the increasing complexity of the energy and environmental challenges we face will inevitably necessitate a more diverse approach. There really is no one magic silver bullet capable of solving all of our problems. Part of the beauty (and complexity) of the current era is that there is so much diversity in technologies, applications, and solutions. This diversity allows us flexibility in adopting the right soultions for each locality to make the best use of local resources in response to that location's specific challenges. Resources can be leveraged in integrating technologies to serve the needs of multiple sectors to solve multiple problems, while providing for value-added markets. For example, we are already seeing the electricity generation and transportation sectors converging; wastes that are an environmental burden are being utilized to create clean energy; and advanced energy storage technologies are enabling wider penetration of clean energy technologies.

Part of the challenge is that the present turning point is multifaceted. On one side, we have a much larger population that has become increasingly more accustomed to the established energy infrastructure and the modern conveniences of this infrastructure. On the other hand, there is a growing sense of urgency, as the environmental challenges our planet faces today are far more pressing than they were before. Moreover, the adoption of renewable technologies at the large scales needed means that we need to shift from an infrastructure that extracts high energy density resources (conventional fossil fuels) from relatively few points of access and delivers them globally, to one where lower energy density fuels (clean energy options) would be collected at low power densities over extensive areas and then concentrated in consumption centers, where the modern urban and industrial areas are expected to consume energy at increasingly higher power densities [2].

When faced with extreme challenges and crisis situations, humankind is forced into action. Most of the environmental laws and regulations put in place can be traced back to a crisis and the accompanying damage that sparked the implementation of such rules. No matter how late, action is worthy. Some damage cannot be fixed. Seemingly harmless consequences that accumulate in the present later confront us as big concerns. To our advantage, we have a significant amount of accumulated scientific data, experience, and insights from lessons learned, all of which help accelerate the pace of technology development and enable faster and smoother energy transitions.

When it comes to new clean energy technologies, one concept entrenched in many minds is that a significant *breakthrough* is required to make these technologies economically competitive. While important advances still need to be made, the more critical component actually required is an adequate *market size.* The growth of the market will drive innovation. The demonstrated feasibility and performance of technologies are important at this point; once technology is proven, prices are driven by the size of the market. Typically, a doubling of installed capacity would mean about a 20 percent reduction in price [3]. Thus, it is crucial to design appropriate policies and provide support to these technologies to help overcome technical, and nontechnical barriers to enable the market to grow.

While *technology development* is an integral component for the successful penetration of clean technologies and our progress toward greater sustainability, it is actually only one part of the "three-legged-stool" required for a balanced future. In fact, it may even be the easiest part of the puzzle to solve. The second leg of the stool is the *consumer.* There have been many examples of new technologies in different areas that have clearly demonstrated the important role the consumer plays. If the consumer is well educated, and motivated, many challenges can be overcome. The consumer may be a family implementing energy efficiency measures in their home; it may be a business converting its vehicle fleet to alternative fuels; or an industry making a commitment to lower their energy use. At any level, an informed, supported, and motivated consumer can have a sizable impact on moving new technologies forward. In making decisions, however, both the consumers and the technology developers need appropriate *policy support,* which is the final leg of the stool. Appropriate policies at the national, regional, or local level may support new technologies in moving down the experience curve, since the market has inherent imperfections. However, setting effective policy is easier said than done, as there are high stakes, competitive politics, and diverse interests involved.

Most importantly, in making any decision we must not forget that the *natural* capital that the planet has provided us is what will sustain us in the longer-term; not the currency-based capital that, while essential, can often also distort the sustainable, comprehensive view that is needed.

References

[1] United Nations Population Fund, *State of the World Population 2011: People and Possibilities in a World of 7 Billion,* New York, NY: United Nations Population Fund, 2011.

[2] Smil, V., *Energy Transitions: History, Requirements*, Prospects, Santa Barbara, CA: Praeger, 2010.

[3] Mallon, K., "Myths, Pitfalls, and Oversights." In *Renewable Energy Policy and Politics,* pp. 5–34, K. Mallon (ed.), London, UK: Earthscan, 2006.

About the Author

Dr. Yüzügüllü started her career with bachelors and masters degrees in environmental engineering, which she supplemented with a doctor of science degree from The George Washington University in environmental and energy management (focusing on hydrogen technologies). This education provides her with a comprehensive, multifaceted *systems level view* of issues related to the way we create, consume, and manage energy. Throughout her career, she has been involved in analyses of a diverse array of clean energy technologies and issues, including techno-economic evaluations, market penetration analyses, technology deployment planning, and decision-making model development for consensus building. Dr. Yüzügüllü's extensive experience and interaction with the federal government, national labs, and industry have provided her with detailed knowledge on critical energy issues, as well as the capability to understand and resolve concerns related to divergent stakeholders in the energy industry. She has also participated in clean-energy-related conferences in various countries, with roles such as keynote speaker, discussion panel moderator, conference organizing committee member, and abstract reviewer. She is currently employed as a senior clean energy analyst with SRA International's Washington D.C. office. She also teaches graduate-level classes on energy management at The George Washington University's Engineering Management and Systems Engineering Department as an adjunct professor.

Index

Recent Artech House Titles in Power Engineering

The Advanced Smart Grid: Edge Power Driving Sustainability, Andres Carvallo and John Cooper

Battery Management Systems for Large Lithium Ion Battery Packs, Davide Andrea

Battery Power Management for Portable Devices, Yevgen Barsukov and Jinrong Qian

Designing Control Loops for Linear and Switching Power Supplies: A Tutorial Guide, Christophe Basso

Electric Systems Operations: Evolving to the Modern Grid, Mani Vadari

Energy Harvesting for Autonomous Systems, Stephen Beeby and Neil White

Power Line Communications in Practice, Xavier Carcelle

Power System State Estimation, Mukhtar Ahmad

Synergies for Sustainable Energy, Elvin Yüzügüllü

For further information on these and other Artech House titles, including previously considered out-of-print books now available through our In-Print-Forever® (IPF®) program, contact:

Artech House
685 Canton Street
Norwood, MA 02062
Phone: 781-769-9750
Fax: 781-769-6334
e-mail: artech@artechhouse.com

Artech House
16 Sussex Street
London SW1V 4RW UK
Phone: +44 (0)20 7596-8750
Fax: +44 (0)20 7630-0166
e-mail: artech-uk@artechhouse.com

Find us on the World Wide Web at: www.artechhouse.com